# RESTLESS CREATURES

# RESTLESS CREATURES

## THE STORY OF LIFE IN TEN MOVEMENTS

MATT WILKINSON

BASIC BOOKS
A Member of the Perseus Books Group
New York

Published by Basic Books,
A Member of the Perseus Books Group



Books published by Basic Books are available at special discounts for bulk purchases in the United States by corporations, institutions, and other organizations. For more information, please contact the Special Markets Department at the Perseus Books Group, 2300 Chestnut Street, Suite 200, Philadelphia, PA 19103, or call (800) 810-4145, ext. 5000, or e-mail special.markets@perseusbooks.com.

Designed by Jack Lenzo

Library of Congress Cataloging-in-Publication Data

Names: Wilkinson, Matt, 1975– , author.
Title: Restless creatures : the story of life in ten movements / Matt Wilkinson.
Description: New York : Basic Books, 2015. | Includes bibliographical references and index.
Identifiers: LCCN 2015037173| ISBN 9780465065721 (hardcover) | ISBN 9780465098699 (ebook)
Subjects: LCSH: Animal locomotion. | Extremities (Anatomy)—Evolution.
Classification: LCC QP301 .W673 2015 | DDC 591.47/9—dc23
LC record available at http://lccn.loc.gov/2015037173

10 9 8 7 6 5 4 3 2 1

For Dad

# *Contents*

# Introduction

> Then God said, "Let us make man in our image, after our likeness... " So God created man in his own image.
>
> —Genesis 1:26-27 (RSV)

That's as much as the Bible has to say about the origin of man and why we are as we are: if we take the above verses literally, God just felt like making a species that looked like him. As explanations go, it's pretty thin—one wonders, for instance, why God happened to be humanoid in the first place. Not that the rest of the living world fares much better. The only vague allusion in the first pages of Genesis to a possible link between an organism's form and its way of life is the description of birds as winged. Such reticence is hardly surprising, given that curiosity is denounced as the mother of all sins a couple of chapters later. Righteousness apparently requires that we take everything for granted. Fortunately, we ended up ignoring this prescription: the first rule of a post-Darwinian, evolutionary worldview is that when it comes to life, we should take absolutely nothing for granted. Living things are as they are largely as a result of the process of adaptation: the gradual accumulation of favorable mutations over countless generations by the action of natural selection.

Now, some would have it that natural selection is therefore the one and only answer to any question about life. But while this claim is sort of true, at least as far as the adaptive features of organisms are concerned, pointing at a single element in a causal chain hardly amounts to an intellectually satisfying explanation. However, we'd be forgiven for thinking that this is about as far as we can go. It's all very well accepting the authority of natural selection as the agent of adaptive change, but the genetic mutations that create the variation upon which natural selection acts occur by chance. Furthermore, the eventual fate of a mutation hinges on the particulars of the environment in which a population happens to find itself. Are we really any the wiser? Is evolutionary history just one damned thing after another—a long sequence of specifics and contingencies? Have we succeeded only in replacing God with Fortune?

Some would argue that this is indeed the case. Ernest Rutherford—widely regarded as the father of nuclear physics—once said that "physics is the only science; all else is stamp collecting." And while he would never have sanctioned such scornful language, evolutionary biologist Stephen Jay Gould was really on the same page when he declared that any rerun of the tape of evolution would produce a living world utterly alien to the one we know. In saying this, Gould was tacitly agreeing that, on the grand scale, evolution is an incomprehensible, unruly beast—opaque to the order-seeking searchlight of science. In this book I offer a wholly different view, for I believe that, contrary to first impressions, there *is* a way of making deeper sense of ourselves and other living things. Life, you see, despite its overwhelming diversity, has a single overriding theme—one that has dominated evolutionary possibility from the very outset. That theme is locomotion—the apparently simple act of moving from one place to another.

It was pterodactyls that showed me the light. These were the animals that attracted my attention when I first became a research zoologist. My choice of subject was partly driven by the usual childhood dreams of dragons and lost worlds not quite snuffed out by a scientific education (thankfully), but it was also a pragmatic one. Flight is not something to be tackled lightly—after all, we humans cracked it only one hundred years ago—so I reasoned that natural selection must have had a particularly tunnel-visioned concern when it came to the pterodactyls. The unforgiving physical demands of flight would surely have dominated their form and behavior completely, as is the case today for bats, birds, and, indeed, man-made aircraft. Such stringent limitations on evolutionary possibility are an absolute godsend for a paleontologist. If I bore these constraints in mind, I thought, then even with limited help from the fossils, and obviously no chance to observe the animals' behavior directly, I could still go a long way toward reanimating the objects of my affection.

I'm happy to report that my faith was justified. With aerodynamics as my guide, even the most basic data became a rich source of information. Although my later work involved virtual reconstructions and wind tunnel tests, the first thing I did was take my chosen species—a magnificent beast called *Anhanguera*—and estimate its weight and wing area from its fossil remains. *Anhanguera*'s wings were enormous—spanning nearly 16.5 feet, with a 15.5-square-foot shadow—but the creature was also surprisingly light, weighing in at a mere 22 pounds, give or take. Now, simple physical considerations dictate that weight must be balanced by the force of lift in steady flight, and aerodynamic theory tells us that the amount of lift depends on wing area and airspeed, to a first approximation. With its big wings and lightweight build, *Anhanguera* could generate enough lift to stay airborne at a remarkably low cruising speed. Which was just as well: those enormous wings weren't suitable for the vigorous flapping that would be required to make up for a substantial airspeed shortfall.

That was just the beginning. The trouble *Anhanguera* had with flapping meant that it needed gravity's help to get up to speed, along with rising warm air currents—thermals—to maintain altitude. It must therefore have roosted on cliffs near tropical seas, whose waters are balmy enough to sustain the thermal fields. In this regard, *Anhanguera* was a lot like today's frigate birds, but the resemblance may have gone further than simple habitat choice. Frigate birds are notorious for their aerial piracy—they plunder fish from other birds on the wing. What many people don't realize is that this objectionable behavior is born of the birds' locomotory "design." Because they need the assistance of gravity to take off, frigate birds cannot risk alighting on the water to feed. Attacking other birds in midair is a perfectly reasonable response to this difficulty. Who knows—maybe *Anhanguera* and its kin were the analogous scourge of the Cretaceous skies.

All that information came from just two numbers: weight and wing area. With a little physical know-how I was able to take some petrified bones and return a fully functional animal, placed more or less securely in the ecology of its time. This experience was to be my epiphany—I would never look at the world in the same way again, for once accustomed to a locomotory point of view I realized that flight is not unique in its power to shape adaptation. On the contrary—I began to see the guiding hand of locomotion everywhere I looked. Thanks to *Anhanguera* I had stumbled upon life's big secret, hiding in plain sight. Locomotion is perceptually immediate—one doesn't need a telescope or microscope to discern it; neither does one have to wait generations to see it in action. It's happening all the time, all around us. My way was clear: I resolved to lay bare the restless heart of the living world. This book is the culmination of my quest.

----------

There are, I think, two reasons for the pervasive influence of locomotion on the design of living things. First, getting from place to place effectively and efficiently is often one of the most important determinants of how many healthy offspring a creature begets, which as far as natural selection is concerned is ultimately the only thing that matters. Survival and reproduction both require that organisms seek out fuel and raw materials for the purposes of growth, repair, and baby-making, while ideally avoiding competitors or any hungry creature that might make a meal out of them. Sexual reproduction often requires that organisms approach each other, and whether sexual or asexual, one's offspring must eventually fly the nest, so to speak, if they're not to become entrenched rivals of their parents or each other. All of which means that locomotion tends to enjoy a high priority in the eyes of natural selection.

The second reason that locomotion leaves such obvious marks on living things is of a more physical nature. It doesn't matter whether we're considering a corkscrewing bacterium, a climbing ape, a sprinting cheetah, a spinning maple fruit, a soaring albatross, a burrowing worm, a swimming swordfish, or a strolling human: everyone without exception must defer to the same underlying physical reality. Organisms are physical objects, after all, and when moving around, they must obey Newton's laws of motion, along with a few other rules and regulations concerning levers, the behavior of fluids, and so on. Given the high selective premium placed on efficient and effective movement, these rules typically impose tight constraints on the shape and behavior of locomotory creatures.

At this point, you might be tempted to employ your own locomotory abilities (metaphorically at least) to run away as fast as possible, but you'll miss out if you do. The laws of which I speak are really not that scary, and the great importance of locomotion to an organism's fitness means that the form and behavior of living things are often attuned to the same physical principles. If there were any doubt as to the truth of this statement, you have only to consider the innumerable cases of convergent evolution that pepper the history of life. Whales and dolphins have been shaped so thoroughly by the demands of efficient underwater locomotion that for a long time they were regarded as fish. The three groups of flying vertebrates—bats, birds, and pterodactyls—have been brought to strikingly similar anatomical destinations by the unbending physical needs of flight. The diversity all rests on one beautifully simple foundation.

An important question may by now have occurred to you. If all organisms are contending with the same physical reality with the same underlying rules, why don't they all look and behave the same? Why is the living world so astonishingly diverse? There are two principal answers to this question, one more intuitive than the other. First, of course, different organisms inhabit different physical environments where (to put it in prosaic terms) the values of certain variables in our equations of motion aren't the same: they may dig through soil, swim through water, fly through air, or move on an interface between these realms, each technique requiring different anatomical and behavioral traits to pull it off effectively. On similar lines, a creature's size has an enormous impact on its physical experience. The largest locomotory entity—the blue whale—is one thousand million million million times bigger than the smallest—a *Mycoplasma* bacterium. The physical changes across so vast a size range have all kinds of locomotory consequences. An elephant can't bound around, for instance, because its legs would need to be impractically thick to withstand the enormous forces involved. A mouse, on the other hand, rarely does anything *but* bound. Air, whose physical presence is barely noticeable to us in most situations, is positively syrupy to the tiniest flying insects, so they can get away with wings that are just tufts of bristles. Such a design is not recommended for a Boeing 747.

The second, less intuitive answer to the diversity question concerns the impact of the past on a creature's present. Evolution is usually a gradual process, for there are strict limits to the extent of viable change from one generation to the next. While big changes are possible (witness the occasional two-headed mutants), they tend to cause catastrophic losses of fitness, and so are quickly eliminated from the gene pool. Future evolutionary pathways are therefore heavily constrained by the present state of an evolving population. By extension, one needs to know a creature's evolutionary past to fully understand why it is as it is now. For no walk of life is this proviso more important than locomotion. Two creatures moving in the same environment may face the same physical challenges, but their adaptive solutions may be wholly different, just because their ancestors came at the problem from different angles. The flying vertebrates again provide a useful illustration. Their wings are superficially similar, but not identical: birds use feathers, bats stretch a skinlike membrane between their elongated fingers, and pterodactyls, despite also carrying membranous wings, supported each with only one finger. These distinct takes on flight in the three groups trace back to differences in their respective ancestors, some subtle, some not so subtle, when they began to test the air.

The history of locomotion may thus be conceived as a 4-billion-year dance between the physical rules of propulsion and the logic of natural selection, with each step dependent on the one that came before. In this book I will retrace this long dance, and in so doing will show how the need to move has shaped the living world. I begin in Chapter 1 with ourselves. Human locomotion is wonderfully amenable to personal exploration and experimentation, making us the ideal platform for learning the ropes of biological propulsion. As a species, we're also surprisingly capable movers by the standards of our close ape relatives. This is something we usually take for granted—when asked to say what makes us special, most people talk about our superior mental faculties. Yet I'd guess that a similar majority would grant far higher status to elite athletes than to Nobel Prize winners. The subconscious high regard in which we hold our locomotory skills is also betrayed by the extent to which the terminology of movement pervades the language of achievement: when we do well, we're "going places," ideas need to "get off the ground," we "chase" goals, "get up to speed" at a new task, make "leaps" of understanding, and "jump at the chance" to try something new. And while we don't often consciously consider the value we place on locomotion, we certainly miss it if it's gone—our locomotory freedom is one of our most prized attributes.

In considering human movement we're going to come face-to-face with our first evolutionary puzzles, the most obvious being our use of two legs rather than the standard mammalian four to get ourselves from place to place. This mystery will serve as a springboard for our journey back through the evolutionary history of locomotion, for it's only by peeling away the layers of recent adaptations that we can hope to find an answer to this curious anomaly. However, in doing so, we're going to uncover yet more locomotory puzzles that will inevitably take us back even further, until we eventually reach the very origin of locomotion itself. Given our interest in understanding ourselves, the focus of our backward trek through time is our own ancestral lineage, but that doesn't mean that the story is ours alone, for the deeper we go, the more widely shared are the ancestors we encounter. That means that the more we learn about our own locomotory past, the more we'll come to understand the broader canvas of the living world, and the more obvious the universal signature of locomotion will become.

To mark the journey down our ancestral lineage I have chosen as way stations a number of key locomotory transitions, each of which forms the central story of a chapter. These shifts in the tempo and meter of the dance of life all granted access to fruitful new ways of moving, and so have special significance in the grand evolutionary narrative of locomotion. In Chapter 2—the first step on our historical journey—we will explore how our tree-dwelling ancestors became two-legged and eventually left the forests behind. Then, in Chapter 3, we will briefly divert from our ancestral lineage to turn our attention to the skies, and to the various origins of the lucky flying animals that now roam therein. In Chapter 4, we will dive beneath the waves, to examine how natural selection for swimming caused the appearance of the vertebrate backbone, before moving on in Chapter 5 to our closer fishy ancestors, who turned their fins into limbs and crawled onto the land. In Chapter 6, as our journey takes us ever deeper, we will learn how the demands of locomotion shaped the fundamental anatomical blueprint of the animal kingdom, with its clearly defined fore-to-aft axis and left/right symmetry. Chapter 7, on the other hand, will look into how animals ended up controlling the locomotory movements of their finely honed bodies, thanks to the origin of the nervous system. Through these major locomotory transitions we will learn how our ancestral lineage was forged and reforged by the demands of movement, and how biological locomotion works in different environments and on different scales.

Many aspects of the transitions I cover on these first six legs of our evolutionary journey are very obviously related to locomotion—such as the origin of flight, of human two-leggedness, and the transformation of fins to limbs. Other changes have a relationship with movement that might seem a little more surprising. Our much-lauded opposable thumbs, for instance, originally had nothing to do with tool use—the digit's realignment was a climbing adaptation. The famous Cambrian explosion—the relatively rapid diversification of animal body types that began about 545 million years ago—was kick-started by crawling adaptations, as we'll see in Chapter 6. Perhaps most striking of all, as discussed in Chapter 7, the brain and sensory organs were originally nothing more than a guidance system—a computer—to coordinate the body's movements to and fro. The evolution of locomotion is about far more than legs, wings, and fins—indeed, the deeper we dig, the more apparent it will become that few if any aspects of an animal's being aren't related in some way to its present or past adaptations for movement.

Chapter 8 takes a different tack, by exploring the various occasions when locomotion was abandoned—or rather, ostensibly abandoned, for we're going to find that the lifestyles of the static owe much to their history of motion. Indeed, I hope to convince you that, strange as it may seem, locomotion has actually dominated the evolution of plants almost as much as it has that of animals. Although plants themselves usually can't move around, their seeds and pollen must for the purposes of sexual reproduction and dispersal. These imperatives have impacted not only on the design of the dispersal agents (witness the helicopter-like fruit of maple trees) but also on the form of the stationary plant that releases them. Height is an obvious dispersal-assisting quality, and flowers are so good at ensuring pollen gets delivered exactly where it needs to go (via insect couriers) that one could regard them as indirect locomotory organs.

The various narrative twists that we're going to encounter on our long journey show that a creature's evolutionary history shouldn't be regarded as a mere straitjacket. Adaptations can open doors to future possibility as well as close them, and this is never more likely than when those adaptations have a locomotory impact. If a creature acquires a new way of moving, through its travels it may end up exposing itself to a whole new set of selective pressures: pressures that might push its descendants in an entirely unexpected evolutionary direction. After all, an organism must experience a new environment before it can adapt to it. Consider flight again: it's a fantastically effective way to get around, but only those creatures that move in the complex world of the forest canopy are ever likely to stumble upon the selective pressures that might eventually make it possible. But this power of locomotion to unlock new ways of living doesn't just apply to the colonization of new environments, as will become very obvious once we embark on the final leg of our journey back in time. In Chapter 9, we're going to see how the adaptive refinement of locomotion in single-celled creatures laid the groundwork for the great multicellular kingdoms that were to come, before we arrive at last at the most important locomotory transition of all: the beginning of locomotion itself. Judged on its evolutionary consequences, this was undoubtedly the most significant transition in the history of life since its origin. Before locomotory powers evolved, life was little more than unusually complex chemistry. Once organisms started to move around, however, they began to encounter each other, opening the doors to predation, parasitism, sex, and symbiosis. In other words, it was thanks to locomotion that life took on its essential character, and it's had the leading role in the unfolding drama of evolution ever since.

The book will end where it began—with ourselves, or rather our mind, for it turns out that we have locomotion to thank for more than our body alone. In Chapter 10, we will see that our curiosity, our joy, even consciousness itself all owe their intangible existence to propulsion. This shaping of our mind by the dance of natural selection and locomotion gives added significance to our search for self-understanding. We've been gifted with an insatiable desire for movement, but this wish has brought us to dangerous territory in recent years, with our locomotory technologies now threatening the health of both our bodies and our minds. Appreciating how we've been built by life's long locomotory dance is therefore no mere academic concern—it may in the end be our best chance of finding a way to live more healthy, meaningful, and fulfilling lives.

Let the journey begin.

1

# Just Put One Foot in Front of the Other

**In which we immerse ourselves in the oft-underestimated magnificence of getting around on our own two feet**

> Know thyself.
>
> —ancient Greek aphorism

I wonder, dear reader, whether you wouldn't mind trying something out. If it's safe and convenient, I'd like you to take a few steps. For more of a challenge, try turning a corner, or should the local lay of the land permit, go up or down a slope or a staircase. By all means, break into a run if you're able and feeling sufficiently energetic. For most of us, unless hampered by injury, disease, or old age, all of this is so easy that we barely need to think about it. If we get the urge to go somewhere, we just go—rarely if ever do we need to devote any attention to the ins and outs of making the journey happen. But in taking our locomotory skills for granted like this, we seriously underappreciate what is in fact a movement machine of dazzling sophistication. Our engineers have built spacecraft that can land on comets, and our computers have beaten grand masters at chess, but we have yet to see a robot whose movements come even close to the elegance, ease, and flexibility of human walking and running.

So, why not try those few steps again, but this time, think about *exactly* what you're doing. How do you initiate and terminate movement? How do you avoid falling over, even when the ground is uneven? How do you turn, or shift up a gear into a run? Why, indeed, switch from walking to running at all? And how are you doing this all with such fantastic fuel efficiency—ten times that of top-of-the-range walking robots, such as Honda's ASIMO?* I realize, of course, that these are difficult questions to answer, for many of the processes that bring about self-propulsion happen beneath the level of our conscious awareness. But we cannot possibly begin our time-traveling, locomotory tour of life on Earth without first turning our enquiring eyes onto our own locomotion: we need to familiarize ourselves with the evolutionary destination before working out how we got here.

## THE SIMPLE ANSWER

For something as commonplace as locomotion, it took us an awfully long time to even begin to understand how it works. Aristotle, a Greek philosopher we'll meet properly in Chapter 4, was one of the first people to thoroughly ponder the problem. His observations and musings led him to conclude that all motions fall into one of two categories. The first—natural motions—are what an object or material does without being forced: what he deemed the "heavy elements," water and earth, naturally fall; whereas the "light elements," air and fire, rise. His second category—violent motions—covered those movements imposed on an object by an applied driving force, which would include locomotion. Take the force away, so he thought, and motion ceases. These ideas accord well with common sense: we need to push or pull a stationary object to shift it, and if we stop manhandling it, the object usually comes to a halt soon afterward.

About two thousand years later, Italian physicist-astronomer-philosopher Galileo realized that there was something deeply wrong with Aristotle's thinking. While Galileo initially accepted that motions could be natural or violent, he reasoned that if an object's natural tendency is to move directly toward the center of the Earth, only movement in the exact opposite direction—that is, upward—could be regarded as purely violent. What about objects moving horizontally? Galileo construed that those motions—directed

* ASIMO stands for "Advanced Step in Innovative Mobility," but the name is also an obvious nod to Isaac Asimov, author of the I, Robot short stories.

neither away from nor toward the planet—must occupy a third category, which he called neutral motions. His great insight was to realize that once external impediments were removed, it would take only a small force to send an object into neutral motion, and once moving, it would take an impediment—friction, for instance—to stop it.

That may sound familiar. Galileo's thoughts, encapsulated in his law of inertia, are essentially identical to the first law of motion formulated by English mathematician-physicist Isaac Newton (1642–1727):

> Every body continues in its state of rest, or of uniform motion in a right line, unless it is compelled to change that state by forces impressed upon it.[†]

That's not to say that Newton's thoughts on movement were simply a rehash of Galileo's. In his second law of motion he extended the concept of inertia, stating that the acceleration (*a*) of an object caused by the application of a force (*F*) is in direct proportion to the magnitude of that force, but in inverse proportion to the object's mass (*m*). In other words, $F = ma$. Furthermore, unlike Galileo, Newton realized that there was no physical basis for granting special status to natural motions: a fall, like any of Aristotle's violent motions, must be caused by an applied force. Taken together, Newton's insights made complete sense of Galileo's famous freefall experiment in which, as legend has it (some maintain that this was only a thought experiment), he dropped two balls of different mass from the Leaning Tower of Pisa. According to Aristotle, the heavier object should have fallen considerably faster, but in fact they struck the ground at nearly the same instant.[‡] The heavier ball's greater mass meant that the force drawing it to the ground was larger, but greater mass also entails greater inertia, so the resulting acceleration was identical to that of the smaller ball. That constant acceleration (roughly 32 feet per second per second at sea level) was the result of the pull of gravity, and is now denoted *g*; the force—an object's weight in the strict sense—can be found by multiplying that figure by the object's mass.

The relationship between force, mass, and acceleration uncovered by Newton is clearly of great importance to locomotion. But where does the

---

† The law has the important qualifier that the observer's frame of reference must not itself be accelerating.

‡ Actually, the heavier ball would have hit the ground slightly sooner, thanks to the relatively strong effect of air resistance on smaller objects, about which we'll learn more in Chapter 3.

force that propels living things from place to place come from? This is where Newton's third and final law of motion enters the picture. The third law is the really famous one, which states that every action has an equal and opposite reaction, but it's also the least intuitive. It isn't immediately obvious that when pushing on an object, the object simultaneously pushes back on you, but with hindsight it's plain that this must be the case. If no force were pushing back, you wouldn't be able to feel the object you were shoving. Similarly, when a falling ball bounces off the floor, there must be a force pointing upward to enable the reversal of direction. Most critically as far as locomotion is concerned, the downward-pointing force of weight that we're all subject to must be balanced by an equal and opposite force directed upward, ultimately derived from the tiny forces acting between the atoms and molecules of the ground, or else we'd sink through the floor. That *ground reaction force* is the key to locomotion. Push back on the ground and it pushes forward on you, accelerating you toward your chosen destination.

Our muscles are responsible for generating the push, though their action is necessarily indirect. That's because muscles can only pull, so skeletal levers are required to convert the motion. The propulsive action of the human leg, for example, is brought about by its extensor muscles. The calf muscles, which attach to the heel bone via the Achilles tendon, swing the foot down and back about the ankle joint; the bulky quads, which run down the front and sides of the thigh to the tibia (the shinbone), straighten the leg at the knee joint; finally, the gluteus maximus (the buttock muscle), and the hamstrings, which run down the back of the thigh, swing the entire leg backward about the hip joint. When the foot is planted on the ground (and as long as there's sufficient friction between the two), the overall backward push of the leg brought about by the contraction of these muscles causes the forward acceleration of the body. That can't continue indefinitely, of course, and because muscles can't actively lengthen, the extensors must be reset by the contraction of their so-called antagonists—the flexor muscles—which each attach on the opposite side of a joint to its corresponding extensor. The principal leg flexors are: the tibialis anterior, which runs along the shin and inserts on top of the instep—this is largely responsible for raising the foot and thereby relengthening the calf muscles; the hamstrings, which bend the knee* and stretch the quads; and the iliopsoas muscles, which run from the lower back and pelvis

* The hamstrings have two jobs because they are biarticular muscles, crossing both the hip and knee joints. Their knee-flexing job is effectively prevented when the leg is bearing weight.

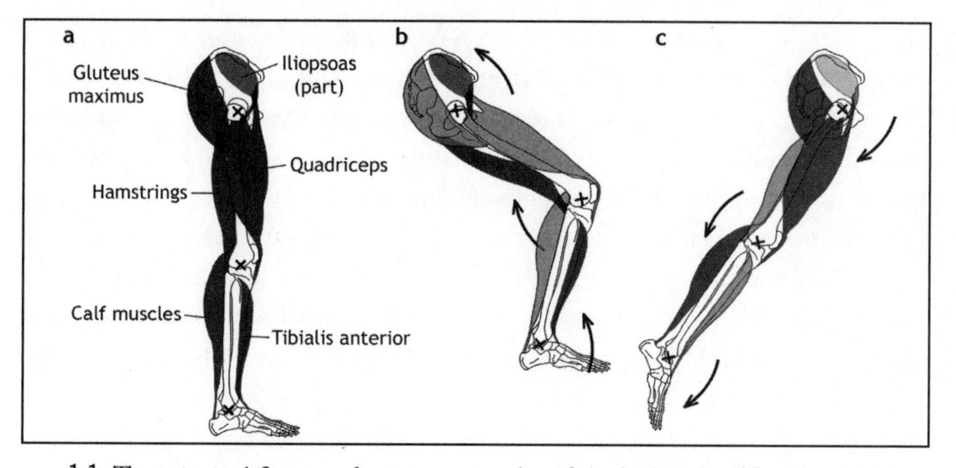


1-1: The principal flexor and extensor muscles of the human leg (a) and a simplified picture of their actions (b, c). The tibialis anterior, hamstring group, and the iliopsoas group (of which only the lower, iliacus muscle is shown here, running from the inside of the pelvis to the femur) flex the ankle, knee, and hip, respectively (b), while the calf muscles, quadriceps group, and the gluteus maximus antagonize these actions by extending the same joints (c). When a muscle is activated (indicated by the darker tone in b, c), its relaxed antagonist (pale tone) is relengthened, thanks to its attachment on the opposite side of the relevant joint. Note that some of these muscles, such as the hamstrings, are biarticular—they cross two joints—and so have alternative actions, depending on the activation state of other muscles or whether the leg is supporting any weight; the hamstrings, for instance, can extend the hip as well as flex the knee.

to the top of the femur (thigh bone)—they pull the leg forward at the hip and stretch the buttock muscles. It goes without saying that the foot must be off the ground during these movements, otherwise we'd push forward and end up back where we started. Fortunately, nature has provided us with a second leg, which can take over the support and propulsion duties during the reset.

So, there we have it—the essential character of walking locomotion, with each leg alternately supporting the body during its *stance* phase and preparing for the next in its *swing* phase. Being a walk, there's no unsupported aerial phase, so both legs are on the ground for at least 50 percent of their respective strides (a stride encompasses one stance and one swing). This *duty factor* can get as high as 70 percent in very slow walking, declining to about 55 percent if we really need to get a move on but can't quite bring ourselves to break into a run.

That's all well and good, but the picture we've painted so far is pretty crude. Nothing we've seen yet couldn't be applied to robots, with servos and motors taking the place of muscles, so we've little indication as to what makes our version so elegant and efficient by comparison. And there's nothing to tell us what makes running different from walking, aside from the reduced duty factor that gives it its characteristic airborne stage. There must be more to it than that. Indeed there is, but uncovering the nuances of human locomotion was never going to be easy. Even the most leisurely of strolls involves moment-by-moment changes in the disposition of our limb segments that happen too fast for even the most dedicated observer to fully grasp, to say nothing of all the behind-the-scenes actions in the body that bring about these movements. What we needed was a way to slow down time and lift the hood on our locomotory engine.

## THE TIME LORDS

The man usually credited for ushering in the modern study of locomotion is the brilliant photographer Eadweard Muybridge. Born Edward Muggeridge near London in 1830, he immigrated as a young man to San Francisco, where he made something of a name for himself as a landscape photographer. His locomotory calling came in 1872, when railroad tycoon and former California governor Leland Stanford invited him to his stock farm in Palo Alto, supposedly to settle a $25,000 bet that a horse periodically becomes airborne when galloping.* Muybridge was at first skeptical that photographic technology would be up to the task, but he gave it a go, and soon showed that, even when trotting, a horse does indeed lift all its hooves off the ground for a split second in each step cycle. His success was the turning point of his life: from that moment, capturing the movement of animals became Muybridge's obsession. He worked intermittently at Palo Alto for several years, where he hatched an ingenious plan. He placed a set of cameras at regular intervals along a track, each rigged so that its shutter was activated by a trip wire stretched across the course. When Stanford's horse Sallie Gardner galloped past, it therefore took a series of photographs of itself. The result was a breathtaking sequence of images that showed for the first time every intricate detail of an animal's locomotory movements. Among other revelations, these pictures proved that the contentious aerial phase occurred not when the legs were at full stretch as many had supposed, but when the forelimbs and hindlimbs were at their closest approach.

---

* Some people believe that the wager detail was added later to spice up the narrative for the press.

1-2: Muybridge's images of Leland Stanford's horse "Sallie Gardner."

Muybridge's horse photographs won him widespread acclaim, and in 1884 he was offered a job at Pennsylvania University to apply his technique to a range of other animals, from baboons to lions. Significantly, he was also asked to photograph human movement, and he duly obliged, producing many sequences of men and women, not just walking and running, but jumping, boxing, somersaulting, dancing, even getting into bed. These works were beautiful and enthralling, and indeed remain so to this day. But in scientific terms, they really only scratched the surface. To work out exactly how we move, much more was needed than a simple series of freeze-frames. Fortunately, at about the same time that Muybridge began to uncover the secrets of horse movement, a Parisian physiologist was busy assembling the tools that would eventually fill in all the details he left out.

Étienne-Jules Marey was the true founding father of the science of locomotion. An exact contemporary of Muybridge (they were born and died mere weeks apart), he became obsessed with the task of, as he put it, translating the language of the body—uncovering and making apparent its moment-by-moment activities. To this end, Marey devised what he called his "graphical method," which could turn all kinds of physiological movements, such as a person's pulse, into readily comprehended readouts, by mechanically transmitting the motion to a stylus, using either a lever or a puff of air in a rubber tube. The stylus inscribed a line on a sheet of smoked paper that, importantly,

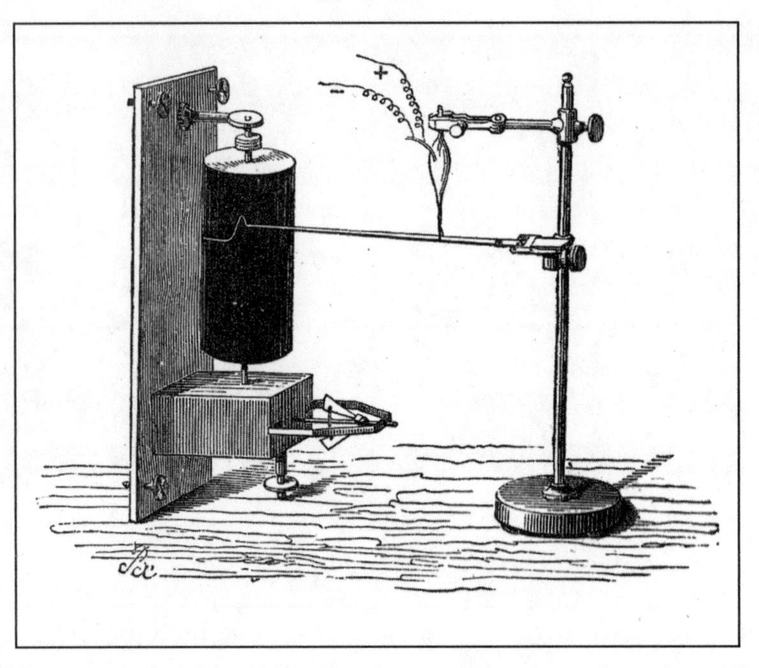


1-3: Marey's graphical method: when the muscle contracts, the stylus is lifted, inscribing a curve on the smoked paper attached to the rotating drum. From *Animal Mechanism* (1874).

was mounted on a steadily rotating cylindrical drum. The resulting readout thus captured the time course of the motion as well as its magnitude.

Marey milked his graphical method for all it was worth (once commenting that to undertake physiological investigations without it was like doing geography without maps) and it wasn't long before his focus shifted from the most minute movements of the human body to the most obvious—the movements of locomotion. For his first project he set out to record the pattern of forces exerted by the feet when walking and running, which he did by mounting customized rubber soles onto his subject's shoes, each containing a small air chamber that communicated with the recording apparatus via the usual rubber tube. Whenever the foot pushed on the ground, a puff of air was sent to the stylus, which deflected in proportion to the magnitude of the ground reaction force. Marey was also interested in how the timing of footfalls corresponded with the rise and fall of the body, so he mounted another device on the subject's head, consisting of a small lump of lead attached to a lever. Given the lead's high inertia, its vertical motions lagged behind those of the person below, so a trace of its relative displacement was presumed to give a fairly accurate representation of the up-and-down movements of the subject.

Marey's technique worked wonderfully, and it wasn't long before he'd extended his analysis to other animals, including horses.* However, while his method told him about certain gross characteristics of an animal's locomotory movements, he could find no way of reliably measuring how fast it or its limbs were moving. Then, in 1879, he saw Muybridge's first photographic sequences and realized with delight that his prayers had been answered. He soon struck up a correspondence, and in 1881 invited the photographer to his home in Paris to give a public demonstration of his image sequences. These Muybridge displayed using his purpose-built zoopraxiscope—a device, made at Marey's suggestion, which projected painted reproductions of his photographs in rapid sequence: it was, in effect, the world's first movie projector. By all accounts, the guests were utterly enthralled—all, that is, except for the renowned painter and horse expert Anton Meissonier, who saw with horror that he'd been getting horse posture wrong for years.

Muybridge's method of slowing down time presented a potential solution to Marey's motion analysis problem, but there was a niggling issue. Because Muybridge's cameras were triggered by the subject, the photographs represented arbitrary points in time, making an accurate calculation of the speed of the body or limbs impossible. Marey realized that if, instead of using multiple cameras, a single photographic plate was exposed multiple times at a known rate, all the necessary information would be available in one image. So, that's what he did. His first multiple-exposure photographs were promising, but if the movements were too slow, the successive freeze-frames were confusingly superimposed. Marey got around this problem, for humans at least, by getting his subjects to dress entirely in black costumes to which reflective strips had been strategically attached. He referred to the technique as geometric chronophotography, but today we'd call it motion capture, and it's still used to great effect by moviemakers when creating convincing CGI characters.

With his moment-by-moment chronophotographic information about the disposition of the limb segments, Marey had at his disposal the means of illuminating human locomotion as never before. Knowing the position of every part of the leg at known intervals of time meant that the velocities and accelerations of each limb element could be calculated; with information

* Thus did Marey discover what Muybridge had with his photographs—that galloping horses took off. In fact, Marey just beat him to it, and many think that the real reason Stanford invited Muybridge to Palo Alto was to confirm what the Frenchman had found.

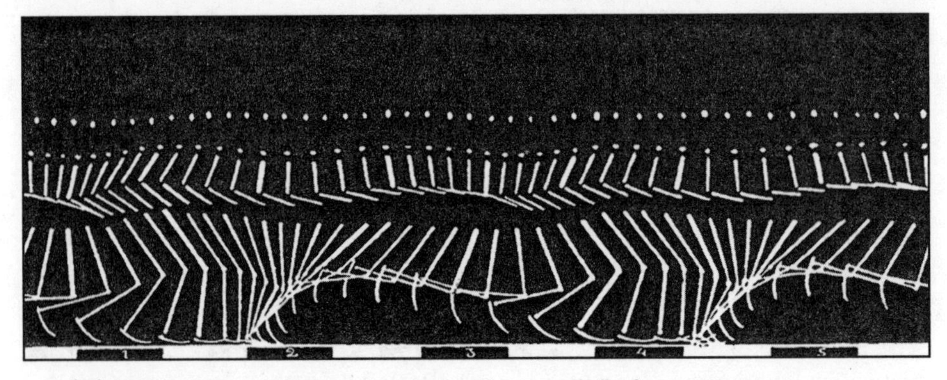

1-4: One of Marey's "geometric chronophotographs" of a running man.

about the masses of those body parts, the muscular forces that created the observed movements could be readily worked out. I say "readily"—in reality this would be an enormously time-consuming task without a computer to help with the calculations, and Marey—now in the autumn of his life—was unwilling to embark on the project. This Herculean labor was left for others. Marey decided instead to spend his remaining years applying his chronophotographic technique to other moving objects—such as flying birds, falling balls, and drifting smoke trails. He did, however, make one further invention that would in time become a vital part of the locomotory toolkit. Back in 1873, Franco-Luxembourgian physicist Gabriel Lippmann had found a way to measure the tiny electric currents generated when a muscle contracted, using a device rather like a mercury thermometer—the mercury rose up a capillary tube when a small current was applied. Marey realized that if a long-exposure photograph were taken of the capillary tube while sliding the photographic plate at a constant speed, the time course of muscle activation could be precisely recorded. He called the technique electromyography (EMG), and though it wasn't applied to the study of locomotion until the 1940s, EMGs are now a standard part of the locomotion scientist's vocabulary.

In the 110 years since Marey's death, the kit has changed beyond recognition: high-speed video has replaced multiple-exposure photographs, and sophisticated electronic transducers now record the ground reaction forces and EMG traces that were once revealed by air chambers and mercury electrodes. But the techniques were dreamed up by Marey, with a little help from Muybridge, and we owe both men a debt of thanks for helping us understand our physical selves more deeply.

## WONDROUS MACHINE

We've learned much since Marey's time. Our quality of life is closely tied to our ability to move around effectively and painlessly, so there's a great deal of scientific interest in the workings of both "normal" and abnormal locomotion. What's more, the sports industry has always been keen to find out how to squeeze the very best performance out of the human body. That level of attention has produced a vast, ever-expanding quantity of information. However, a single theme unites much of what we've found out over the last century—one that will recur again and again in our tour of the evolutionary history of locomotion. It's something I alluded to earlier: compared to even the best humanoid robots, our walking and running techniques are astonishingly cheap. How is it that we're able to move from place to place using a mere tenth of the energy expended by ASIMO?

As far as walking is concerned, the basic answer to this question has been suspected for a long time. As we amble along, our center of mass—the average position of our weight distribution—repeatedly arcs over the supporting leg, and so bears more than a passing mechanical resemblance to a pendulum (albeit inverted, like an old-fashioned metronome). A pendulum requires no force input once set in motion—it keeps swinging back and forth thanks to a perpetual interchange of energy between two forms: kinetic energy and gravitational potential energy. The gravitational potential energy of a pendulum bob depends on its height, so it reaches a maximum when the bob is at either end of its arc. With nothing keeping the bob up there, it then accelerates downward under the influence of gravity, so its potential energy is converted into kinetic energy—the energy borne by any moving object. This reaches its maximum at the bottom of the arc (when the bob is moving fastest), whereupon it is reconverted into potential energy as the bob climbs again. When the kinetic energy is used up, the bob reverses direction and falls back down. That interchange of kinetic and potential energies is exactly what happens to us when we walk. In effect, we intentionally start to topple forward on every step, and rely on the swing leg getting into position in time so that we can use the kinetic energy of the fall to lift our center of mass again (and save ourselves from ending up flat on our face), ready for another fall.

This pendulum-like interchange means that our leg muscles don't have much to do when we're walking, as you can see in Figure 1-5, which shows EMG traces of selected hip, knee, and foot flexors and extensors during a typical step cycle. Only the calf muscles and the lowly tibialis anterior give

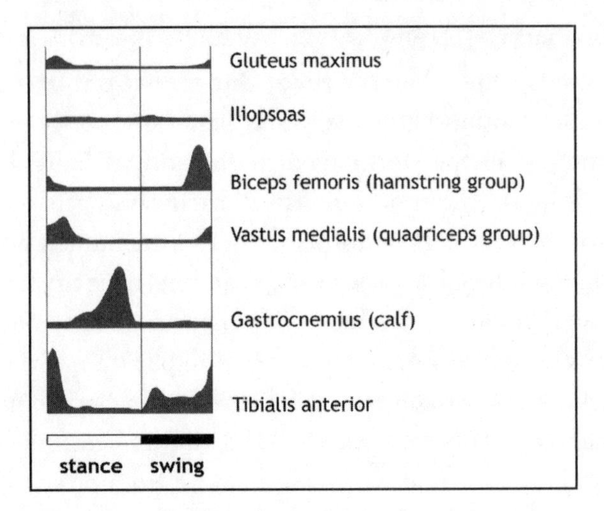


1-5: Electromyograms showing the activity of representative extensor and flexor muscles of the human leg during walking.

even a semblance of hard work. The calf muscles contract to prevent the ankle's collapse during the leg's stance phase, and also give a push prior to liftoff (or toe-off, to use the technical term) to compensate for the inevitable braking effect of the opposite leg's heel-strike. The tibialis anterior has the seemingly insignificant job of flexing the ankle to raise the foot: it's active at the beginning of the stance phase and the entirety of the swing phase (we'll see why later). The bulky hip and knee extensors are required to give only little nudges, and the hip flexors barely do anything at all: the potential energy gained by the leg during the latter half of its stance phase is enough to get it swinging forward passively. In effect, the swing leg acts as a typical pendulum while the rest of the body is busy being an inverted pendulum.

Perhaps the most surprising observation is the lack of knee flexor activity at the beginning of the swing phase, for we know the leg bends at this point—it has to if it's to clear the ground. This knee flexion occurs automatically, thanks to the slowing of the thigh toward the end of the stance phase: as long as toe-off happens at the right time, the lower leg keeps going, and the leg bends. If we walk too slowly, this doesn't quite work: the reduced speed of the swinging leg means that the knee doesn't flex enough, and if the hamstrings aren't activated to compensate, the foot scuffs the ground (the sullen teenager gait). Conversely, if we walk too fast, the intrinsic pendulum action of the leg isn't quick enough to save us from a fall, so once again, extra muscular work is required. Try it and see how it feels. Moving cheaply means moving at the speed at which the body wants to move.

Robot designers are beginning to cotton on to the economic benefits that could be reaped by exploiting a walking body's intrinsic dynamics, thanks mainly to the pioneering work of Tad McGeer at Simon Fraser University, British Columbia, in the 1980s. McGeer designed a number of "passive walkers" that could saunter down a gentle slope with zero power input, and some contemporary robot designers are now looking to these models for inspiration. Engineers at Delft University of Technology, the Netherlands, have built a range of "pumped" passive walkers, bearing such colorful names as Denise and Flame, with a bit of actuation here and there so that they're not always confined to walking downhill. They may not look as polished as the vastly more expensive ASIMO and its ilk, but their motion looks much more lifelike, despite (or really because of) their limited control.

Useful though it is, the pendulum-like movement of our locomotory apparatus isn't perfect, because the kinetic/potential energy interchanges aren't 100 percent efficient. Some energy is always lost as heat—mostly when we come crashing to the ground at heel-strike, hence the need for some extensor muscle activity at this time. The energy loss isn't as bad as it might have been, however, because we limit the up-and-down bobbing of our center of mass as much as possible by dropping the hip of the swing leg. This lowers the center of mass a little at the highest point of its excursion and gives us a smoother ride.* As usual, it happens automatically: supporting the body on one leg leaves the center of mass dangling on the inside, and a tilt of the hips naturally ensues. We don't usually have to worry about toppling because our thighs are designed to angle inward: a slight hook to the end of the femur ensures that the lower legs thence make their way straight down to the ground. This is called the valgus knee morphology, after the Latin for "knock-kneed," and I mention it now because, being a skeletal feature, it's the sort of thing that can give us clues about how our extinct ancestors walked—more on this in Chapter 2.

As is the case for so many activities, the beneficial effect of our automatic hip tilt only works if used in moderation. If left entirely to its own devices, the hip drops too far, causing us to walk with what's known as a Trendelenburg gait: to keep the center of mass over the supporting foot rather than way over to the inside (which would cause a fall), we either have to lurch the upper body from one leg to the other or bend our spine from side to side like catwalk models. Both countermeasures cost energy. Trendelenburg gaits are prevented

* Bending the knees midstance would also do the trick, but requires so much additional muscular effort that it defeats the object of the exercise.

mainly by the gluteus medius muscles,* which run from the top of the femur to the hip bones. If you put your hands just under your hips while walking, you should be able to feel these muscles firing during each leg's stance phase.

Our final piece of energy-saving wizardry is arguably the most complicated. The human foot, containing twenty-six bones wrapped in a spider's web of ligaments and tendons, is a marvel of bioengineering. In every step it must fulfil three different jobs: it has to absorb the shock of heel-strike, then provide a stable platform for the stance leg as the body arcs overhead, before finally acting as a lever to execute the push at toe-off. These may not sound like unreasonable demands, but the anatomical requirements for the tasks are irritatingly distinct: the first two need the foot to have a degree of give, to dampen the peak impact force and then mold to the ground, but the last needs the foot to be as rigid as possible. You might think that the design would therefore represent some kind of jack-of-all-trades compromise, but remarkably, evolution has found a way to enable a functional transformation throughout the stance, allowing us to get the best of both worlds.

The foot's split personality rests in the fact that it has two subtly different configurations: a loose-jointed, flexible state and a compacted, rigid state. Once again, a passive mechanism is responsible for switching between the two. It's called, rather quaintly, the windlass mechanism, after the ancient trick of hoisting heavy loads by wrapping the suspending cable or rope around a crank-operated drum. The "cable" of the foot is a tough fan of connective tissue—the plantar fascia—that spans the sole from the heel to the far ends of the metatarsals (the long bones that support the toes). These metatarsal heads, particularly the one at the base of the big toe, act as the drum. Now, for most of the stance the foot is in its neutral, compliant state to best fulfil its shock-absorbing/molding duties. Toward the end of this phase, however, as the heel is lifted by the calf muscles, the toes are bent back, causing the plantar fascia to wrap around the metatarsal heads and stretch, pulling the end of the heel bone forward. Sticking with the windlass analogy, the calf muscles thus provide the crank action as a by-product of their extensor function. Thanks to the configuration of the various bones that connect the heel bone to the metatarsals, that motion locks the foot bones together, converting them into a rigid lever in preparation for toe-off.

* The gluteus medius muscles also have the important job of setting us off. When they contract, they give a little sideways push to the ground, which shifts the weight to the opposite leg. Then it's up to the tibialis anterior to pull the body forward a bit. Gravity does the rest.

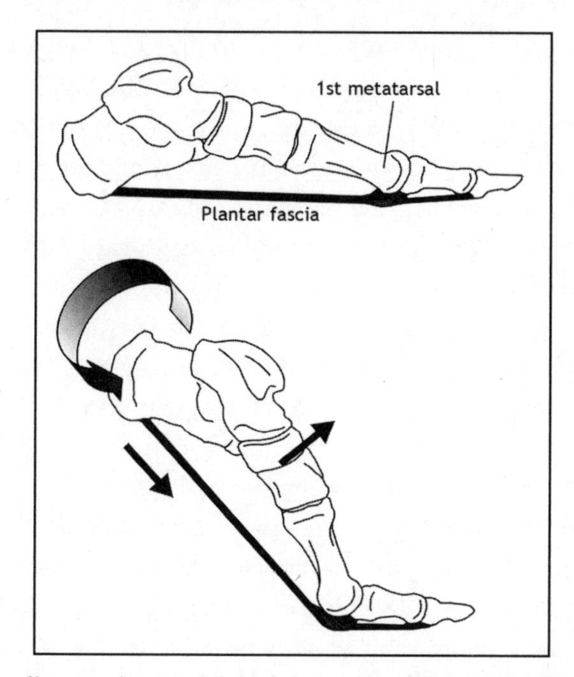


1-6: The windlass mechanism of the human foot. When the heel is raised, the plantar fascia tightens as it wraps around the head of the first metatarsal, pulling the heel bone forward, raising the instep arch, and twisting the foot about its long axis. These actions lock the foot, converting it into a rigid lever prior to toe-off.

Given the extraordinary functional complexity of our feet, it might now seem a shame that the cultural guidelines of our current civilized existence cajole us into encasing our amazing appendages in crude, monofunctional shoes, thus suppressing their intricate workings. This is arguably no big deal when we're walking, as the forces to which the feet are subjected are relatively mild. Running, however, is a different story, and many are beginning to wonder whether Tolkien's hobbits have the right idea, and that we'd get a lot more out of our locomotory machinery if we went around barefoot. I'll come back to this point later. First, we need to take a closer look at the high-performance end of human locomotion.

## CHANGING GEAR

Walking is a fairly forgiving mode of progression, when it comes to speed range. We can shuffle at a snail's pace or motor along at over 5 miles per hour (about 7.5 feet per second), although about 3 miles per hour (4.5 feet per second) is our typical preferred speed as adults. This is the speed at which we exploit the inherent dynamics of our body to maximum effect, returning the most miles per gallon, so to speak. However, in the big, bad, dog-eat-dog world in which we evolved, where sluggishness can result in starvation or the clamping of a predator's jaws around your neck, that 5 miles per hour begins to look woefully inadequate. The partnership between natural selection and physics has therefore provided us with a second gear—running. However, it isn't immediately obvious why this should be necessary. Why can't we just keep walking faster and faster? After all, this is basically how speed is increased on wheels—their revs per minute keep going higher and higher (ignoring the complications of gearing the power transmission).

For a legged animal, speed can be increased in two ways: by extending the stride length (the distance between successive footfalls) or increasing the stride frequency (the number of steps in a given amount of time), or both. When walking, with its proviso that at least one foot must be in contact with the ground at all times, stride length is heavily constrained by the length of the legs. One could conceivably push it to twice the leg length, by sinking into splits on every heel-strike, but this would entail such outrageous vertical oscillations of the center of mass that it's quickly ruled out on energetic (not to mention safety) grounds. In practical terms, then, only a modest increase in stride length is possible when walking. What about stride rate? This, again, can be increased a bit, but as we saw earlier, it will cost us to push the legs too far beyond their intrinsic pendulum dynamics.

There's a more insidious issue at work here. No matter how you go about it, it's physically impossible to walk beyond a certain speed: for a typical adult male, the upper limit is just under 10 feet per second—just over 6.5 miles per hour. The problem stems from the universal tendency of all moving objects to keep going in straight lines. Following a circular path, or part of one, isn't natural—it requires the imposition of a force perpendicular to the direction of motion. Called the centripetal (center-seeking) force, it causes an object to accelerate toward the axis of the resulting curved path. Consider an orbiting spacecraft. It dearly wants to shoot off into deep space (according to Newton's first law), but gravity provides a centripetal force that reins it in and keeps it going around and around. In effect, the craft is constantly falling toward the Earth, but because the planet itself is round, the vessel never approaches the surface.* If we want to get quantitative, the centripetal acceleration is simply

* It's this perpetual plummeting that gives astronauts the illusion of weightlessness: in reality, being only a couple of hundred miles (on average) above the Earth's surface, they have almost the same weight as they had on the ground.

the square of a circling object's velocity, divided by the radius of its circular path: $a_c = v^2 \div r$. The centripetal force is found by multiplying that by the object's mass (remember, $F = ma$).

So, what's all this got to do with walking? Well, in every stance phase the center of mass describes a curved path as it vaults over the supporting leg, with a radius of curvature roughly equal to the length of said leg. Gravity supplies the necessary centripetal acceleration: a constant 32 feet per second per second, give or take. If the forward velocity increases, however, there will come a point at which the acceleration due to gravity is no longer sufficient to keep the radius of curvature down, and we take off. At this point, of course, we won't be walking anymore. With an average adult male leg length of nearly 3 feet, the velocity limit is the square root of (3 × 32) = just under 10 feet per second, QED. Now, we can cheat this a little bit. Accentuating the hip drop increases the radius of curvature of the center of mass's path and diminishes the centripetal force requirement at high speeds, which is what gives racewalkers their curious wiggling gait; they also use fore-and-aft hip swings to raise each leg's axis of rotation and effectively increase the length of the legs. The technique is energetically expensive but effective—at the time of writing, the world record racewalking speed (held by the Russian Mikhail Shchennikov) is an incredible 15 feet per second—that's over 10 miles per hour.

World champion or not, however, all of us eventually hit the walking speed limit, and if we want to go any faster, we must change gear. Running is distinguished from walking not just by the unsupported aerial phase of the step cycle, but by the relationship between the body's kinetic and gravitational potential energies in any given moment. Remember that when walking, our energy alternates between these two forms. That's not the case in running. The center of mass undergoes the same deceleration as it passes over the supporting foot, but it drops in height at the same time, owing to the pronounced flexion of the knee at impact (an important shock-absorbing measure). The changes in kinetic and potential energy are thus in sync—there's no interchange.

That sounds like bad news. If kinetic energy decreases without a simultaneous gain in potential energy, one would think it must be dissipated wastefully as heat on impact, with the muscles then having to supply the shortfall prior to takeoff. That this isn't entirely the case was discovered when the metabolic rate of runners was measured in the 1960s, and found to be only 65 percent of what would have been needed to account for the presumed reinjections of energy. That could only mean that something else (other than height) was storing the energy of impact. It didn't take long to figure out what was going on. Our tendons are not merely connecting cables linking muscle to bone. They can also function as springs, storing some of the kinetic energy used to stretch them as elastic energy, which is reconverted to kinetic energy when the tendon is released. In running, the Achilles tendon is the prime storage site. As the ankle flexes on impact, this tendon is stretched and energized; that energy is then used to help extend the ankle at toe-off, supplementing the work of the calf muscles. If running barefoot, the plantar fascia and the ligaments of the arch of the foot act in the same way: the arch flattens slightly during the stance phase and springs back into shape near toe-off.

This all sounds wonderful, and indeed it is, but it's worth noting that using tendons for elastic energy storage comes at a price. Tendons aren't simply elastic bands linking two bones—there's always a muscle at one end, and that muscle has to be in a contracted state for elastic energy storage to work. The Achilles tendon, for example, can only function as a spring if the calf muscles are activated. Despite this hidden cost, elastic energy storage is vital at our highest speeds. In fact, even at the low-speed end of running—below the maximum walking speed—elastic interchange was once thought to be an even better energy-saving trick than the pendulum mechanics of our more sedate gait. That would explain why we tend to change gear at around 6.5 feet per second, which is well below the walking speed limit. Or at least, it would explain it if it were true. Recent experiments show that we do ourselves no energetic favors by switching to running at such a low speed, which makes one wonder why we do it. Oddly enough, given that this is such a commonplace event, we're still not absolutely sure. One contributing factor seems to be perceived stress in the tibialis anterior (the main ankle flexor), which, if you remember, is one of the hardest-working muscles during walking. It has two main functions: it fires at heel-strike to ensure that the foot doesn't slap down on the ground, and then again at the beginning of the swing phase to keep the toes from dragging in the dirt. It seems that at the highest walking speeds it all starts to get a bit much for this little muscle, and we switch to running to give it an easier time. If artificial ankle flexor assistance is provided (running a strap from the forefoot to the knee does the trick), we do indeed tend to delay the transition to running.

Regardless of why we switch to running exactly when we do, the overall benefits of our second gear are obvious: thanks to its aerial phase, running does wonders for our stride length (it can go up to 11.5 feet, although 6.5 is typical), and can potentially push the maximum speed up to just over 34 feet per second, or 23 miles per hour (if you're Usain Bolt*). In mechanical terms

* At the time of writing, Jamaican sprinter Usain Bolt holds the world records for both the 100-meter (9.58 seconds) and the 200-meter (19.19 seconds) races.

it's a surprisingly easy switch. The muscle-firing sequence used in running is almost identical to that used in walking, although the duration and intensity of contractions are ramped up as you'd expect, particularly in the hip muscles. The only major sequence difference in running involves the timing of activation of the calf muscles: they fire much earlier in the step cycle, such that they're active at touchdown, to enable the spring-loading of the Achilles tendon. If running on the balls of the feet, this early activation also allows the calf muscles, rather than the heel pad, to absorb the larger impact forces of landing. This extra damping becomes increasingly important at our highest running speeds, thanks to the low duty factor (more time in the air, less on the ground) and correspondingly higher peak forces: the time-averaged ground reaction force must still balance body weight, regardless of the fact that the force is now applied to the body for a much shorter period of time.

Evidence is accumulating that relying on the heel's shock-absorbing function at all while running is really rather bad for us. Habitually barefoot runners generally avoid heel-strikes, for the simple reason that they hurt. Of course, nice thickly soled, air-cushioned shoes eliminate the pain, but might they also be lulling us into a false sense of security, causing us to pass dreadful shocks up our legs and spine without our being aware of them? Experiments conducted by evolutionary anthropologist and barefoot runner Dan Lieberman and his colleagues at Harvard indicate that this may indeed be the case. It wasn't that long ago in geological terms that our ancestors all wandered around barefoot, and even in more recent times (up to the 1970s) we generally ran wearing thin-soled, minimal shoes, such as sandals or moccasins, which had little mechanical impact. We might therefore conclude that our locomotory machinery was never "intended" to work with chunky, high-tech sports shoes stuck to the ends, and that we should ditch these unwieldy accessories at our earliest convenience. Indeed, running injuries are alarmingly commonplace these days, hinting that the body is too often subjected to excessive stresses. That said, we pampered Westerners of current generations have generally grown up in the company of such shoes, and would be expected to have adapted to them (in the nonevolutionary sense) over the courses of our lives. In support of this notion, newbie barefoot runners sometimes sustain injuries because their feet and calves aren't strong enough to cope with the new regimen, although the right training program may eliminate such transition problems. We clearly need to do more work before we can make any definitive pronouncements on the to-shoe-or-not-to-shoe question; hopefully we'll have a better idea in a few years' time.

Before moving on, it's worth noting that walking and running are not our only locomotory gears: we can also skip. Skipping is a more complicated gait than either walking or running, for it combines the potential/kinetic energy interchange of the former with the elastic energy storage of the latter—in fact, its energetic profile is effectively that of a gallop, but with two legs rather than the standard four. For all that, though, it's our least efficient gear, partly because the vertical oscillations of the center of mass as we spring up and down are more extreme than in walking or running. Along with the fact that skipping around tends to raise the eyebrows of passersby, that would explain why we don't use this gait much, at least as adults (children are often rather fond of skipping, although we don't yet know why). But it does have its merits in some situations. In the low-gravity environment of the Moon, for instance, skipping is far and away the most effective mode of progression. Walking doesn't work so well here because the usual efficient interchanges of kinetic and gravitational potential energy rely on the earthly relationship between mass and weight (our mass on the Moon is unchanged—it's only our weight that goes down). Running is even worse: low weight means low friction, which brings a high risk of slippage. In a skip, this risk is significantly reduced because the legs don't stray so far from vertical at touchdown and toe-off. Skipping has its uses on Earth, too. For example, many people (myself included) skip to go down stairs quickly, and while this phenomenon has never been scientifically examined (as far as I'm aware), you might be able to think of some reasons why this is better than running. I'll leave you to ponder this little mystery at your leisure. Please don't break your neck in the process!

## NATURAL-BORN RUNNERS?

I hope you can now appreciate how extraordinarily well we humans have been "designed" for locomotion. From hip to toe, there's barely a single feature of our lower extremities and the way we use them that can't be explained in the light of propulsion. Whether it's the shape and size of our joints or the timing of our muscle contractions: all seem wholly dedicated to getting us from place to place as quickly, safely, and economically as possible. Not that this is an entirely waist-down phenomenon. Every time a leg swings forward, it imposes a twisting motion on the rest of the body that would require costly correction if it were not for an equivalent swing of the opposite arm (which, by the way, is a mostly passive motion). That itself requires a flexible waist to enable the shoulders and hips to twist in opposite directions, so it's a good thing our rib cage doesn't extend all the way to our pelvis. But we don't want the twisting motion to be transmitted to the head, otherwise our eye line would be forever veering left and right: we therefore have a comparatively thin neck and finely honed postural reflexes to help us maintain a steady gaze. And the torso can't be too floppy or else it would lurch forward at every footfall, thanks to its inertia, putting us in danger of falling on our face. This unwanted motion is prevented by our erector spinae muscles, which run up either side of the spine from the pelvis and contract on every step to give the torso a little backward tug (put your hands on the small of your back when walking, and you'll feel them tense).

These many locomotory adaptations leave us in little doubt as to the historical importance of movement to our ancestors. If our biology makes us good movers now, that can only be because, time and time again in the distant past, those people who had the luck to bear genetic mutations that improved their locomotory prowess were the ones who, on balance, had better survival rates and/or more children. Their favorably mutated genes and the skilled movement that came with them therefore spread through human populations at the expense of their less fit genetic counterparts. Such are the implacable ways of natural selection. Working out *why* locomotion was so important to our ancestors, however, is another matter. At first it might seem obvious that being able to move faster would make successful escapes from ravenous predators more likely—until you realize just what our ancestors were up against. Our top athletes today can maintain a speed of a little over 20 miles per hour for a few seconds. On the savannah of our ancestral African home we were pitted against the likes of hyenas and lions, who can double those speeds. Of course, there's the old adage that you don't need to run faster than the predator, just faster than your friend; but even so, we can be pretty sure that it was their strength in numbers that saved our ancestors from death by carnivore, not their pace.

If the answer to our question can't be found up the food chain, then perhaps we need to look down, in the direction of potential prey. Here again, though, we're looking at creatures like deer and antelope that can outperform us easily in sprint chases, with considerably higher maximum speeds that can be maintained for minutes, not seconds. The high-velocity end of things is beginning to look decidedly unpromising. But what of the more sedate, walking end of the spectrum? Could fuel economy have been the primary target of natural selection? As it happens, our energy expenditure when walking is unusually low for a creature of our size. While perhaps less glamorous than the game of life and death played out by today's savannah predators and their ungulate prey, an ability to travel long distances cheaply may be no less important in selective terms, especially in such open environments, where a decent meal may be a fair hike away. After all, the less energy expended getting from breakfast to dinner, the more will be available for the pressing business of reproduction.

Something's not right here, though. As Dennis Bramble of the University of Utah and Dan Lieberman have pointed out, too many aspects of our locomotory anatomy just don't fit with any mode of progression other than running. The energy-storing role of our well-developed Achilles tendon is barely used when walking; the same can be said of the spring action of the arch of the foot. Our swinging arm technique, though of some use in a stroll, is much more important in a run, when the rotational disturbance set up by the now violently swinging leg can't be balanced by the hip muscles of the opposite leg because said leg is in midair. Finally, our buttocks are too big. Our gluteus maximus muscles are relatively enormous compared to those of our ape cousins, but they barely tick over when walking. Only when we run do they come into their own, giving each leg a powerful kick at toe-off and assisting the erector spinae muscles in stabilizing the torso.

Notwithstanding all that, it can't be denied that we're pretty terrible sprinters in the grand scheme of things. But there are other ways to run. It's when we choose not to go flat out that we shine. Today's top long-distance runners can maintain speeds of 14.5 miles per hour, and even non-professionals can manage between 7 and 9.5. That may not sound like much, but with a decent level of aerobic fitness (which our ancestors surely had), these speeds can be sustained for hours. When assessed over such spans of time, our running performance begins to look quite respectable when compared with potential mammalian prey (it goes without saying that endurance running would have done little good when pitted against a sprinting predator). As long as the target can be tracked, even antelope can be outpaced by sufficiently determined humans. However, catching up to prey is one thing—it's quite another to dispatch it without nature's usual tools of tooth and claw. Sharp-edged stone tools might be thought to have fit the bill, and these were in use by the time endurance running adaptations appeared in our lineage. Our earliest known fossil relative to have had such traits was *Homo ergaster*, a rangy species (adult males may have been 6 foot 2 or more in height) that lived in eastern and southern Africa between 1.8 and 1.3 million years ago: for comparison, the earliest stone tools are about 3.4 million years old. However, while these stone implements could have been used to butcher carcasses, as offensive weapons they leave much to be desired. They work better when attached to spears, but these didn't come on the scene until a few hundred thousand years ago.

If active hunting wasn't possible, perhaps scavenging for meat presented a viable alternative? Endurance running could have provided an advantage here, too. Savannah scavengers typically use circling vultures as a cue to a kill (or a death by natural causes), and from then it's a race against time to get there before the carcass is scoffed. Recent mathematical models of day-to-day life on the savannah made by Graeme Ruxton and David Wilkinson of St. Andrews and Liverpool Universities indicate that thirty minutes is probably the critical time window—if you take any longer to get to the meat, it won't be worth your while. Unfortunately, this probably isn't long enough for humanoid endurance running to make enough of a difference: we can only win longer races. So, where does this leave us? If neither active hunting nor scavenging provided sufficient selective impetus for the evolution of our endurance-running adaptations, what did? Well, there is a third option: keep harrying an animal for long enough and it might eventually die of exhaustion or heatstroke. The trick is obviously to not die of exhaustion and heatstroke yourself, and here we see a potential role for another couple of strange human features. We are almost unique in our use of sweat to cool down. The vast majority of mammals can't do this because the oils in their pelt would impede the passage of water,* but this is no problem for us because our body is practically hairless. The importance of this physiological trick is not to be underestimated. Our energy efficiency when running is poor by mammalian standards—even in endurance mode we use 50 percent more energy to go a given distance than does a typical four-legged mammal of the same mass. Yet *they* are the ones that run themselves to death, thanks to our vastly more effective coolant system.

Just to drop a final fly in the ointment, one important part of our locomotory design doesn't fit with running. Most terrestrial mammals of a more athletic persuasion permanently stand and move around on tiptoe (giving rise to the myth that horses' knees bend the opposite way to ours—that's the ankle you're looking at). Birds are much the same. The reason is simple—this is an easy way to extend the stride length and increase speed. Many lineages have continued this trend by elongating the feet. Our stubborn insistence on using

---

* Horses are an exception—they use sweat to cool, while also secreting a detergent (the protein latherin) to break up the oil.

the whole sole would be bizarre if we really were dedicated runners. But, as Christopher Cunningham of the University of Utah and his colleagues recently found using treadmill experiments, our postural preference makes complete sense from a walking perspective. When their volunteers walked on tiptoe, they used over 50 percent more energy than when walking normally (by all means, try it yourself). By contrast, however, switching posture made no difference to their energy expenditure when running. When all's said and done, arguing over whether we're chiefly adapted for running or for walking may therefore ultimately prove futile: natural selection seems to have been concerned with both gears.

It is of course difficult to be completely sure of the selective factors that caused our specific suite of locomotory adaptations. What we can say is that much of our present form is attributable to the way we were selected to move, even if some of the details of that relationship have yet to be ironed out. And our form is just the start. Whether our ancestors were hunters or scavengers, the large amount of meat in their diet was only possible because they had a locomotory system that could get hold of it. This makes us unique among our ape cousins, for apes subsist almost entirely on fruit, nuts, and leaves (even the relatively bloodthirsty chimps take only the occasional monkey as a treat). The significance of this dietary shift cannot be overstated. With meat on the menu, the nutritional quality of our food shot up. No longer did we need the lengthy guts of our vegetarian ancestors, so all the resources that were once needed for the development and maintenance of capacious intestines were available to be used by other organs—mainly the expensive brain, which about 2 million years ago began to increase in size exponentially. We are what we eat indeed, and because our ancestors could only eat what they could catch, it's no exaggeration to say that it was our locomotion that made us what we are.

----------

It would be easy to think that we now have the measure of ourselves. However, it takes only a quick glance at our relatives to see that when it comes to explaining why we are the way we are, we've barely scratched the surface. Many of our mammalian cousins have been driven by the same locomotory selective pressures that guided us, yet have ended up at radically different evolutionary destinations: most obviously, we do our moving on two legs, while the champion nonhuman runners do theirs on four. As a result, on top of their second gear—usually a trot, with the left/right alternation of the forelimbs occurring simultaneously with the right/left alternation of the hindlimbs—they can really put the pedal to the metal and switch to galloping. With its aerial phase between the footfalls of the hind- and forelimbs, galloping (unlike the otherwise similar human skipping) adds the flexibility of the back to the locomotory equation, allowing some mammals to achieve truly phenomenal speeds. Cheetahs—the champion terrestrial sprinters—bend their back so much when they run that their stride length can reach 23 feet or more. As a result, they can pelt along at 60 miles per hour. Why did our ancestors write off this fantastic high-performance option by going bipedal? To find the answer to this question, we're going to have to go back much further than a mere 2 million years. The time has come to meet our closest relatives, who are going to show us the momentous consequences of hanging about in trees.

2

# Two Legs Good

## In which we discover how our tree-clambering ancestry made us who we are

Two of far nobler shape erect and tall
Godlike erect, with native Honour clad
In naked Majestie seemd Lords of all
And worthie seemd, for in thir looks Divine
The image of thir glorious Maker shon

—John Milton, *Paradise Lost*

Ask anyone to think of an image that sums up the process of evolution (as I, dear reader, am now asking you), and it's a fair bet that most will call to mind *The March of Progress* or one of its many successors. This picture, created by natural history artist Rudolph Zallinger in 1965 for the book *Early Man* by F. Clark Howell, depicts fourteen extinct relatives of modern man variously shambling, trudging, and striding from apelike left to humanoid right. It perfectly captures the idea of an inexorable evolutionary improvement from primitive ape through ape-man intermediates to our current manifestation—and is, as such, utterly misleading. Evolution doesn't work like this. *Homo sapiens* represents just one of countless experiments in the art of living, not the shining apotheosis of 4 billion years of adaptive change: the tree of life is more shrub than redwood. In fairness, neither Zallinger nor Howell meant to imply otherwise (notwithstanding the ill-chosen title), and many of our apparent forerunners in the illustration are correctly acknowledged in the text as offshoots of our ancestral line—x-hundred-thousandth cousins many times removed, not great-great-great-etc. grandparents—but this message is easily swamped by the linear imagery, and by one particularly conspicuous aspect of the implied evolutionary trend. The third member of the time troupe, labeled *Dryopithecus* (an extinct ape that lived between 12 and 9 million years ago), has been modeled rather closely on a knuckle-walking chimpanzee, and its presence is key to the *March*'s iconic status. In the many subsequent versions of the picture, the indispensable common factor is the postural shift from a hunched, quadrupedal, chimplike ape to a fully upright, bipedal human.

This idea of the literal rise of humanity from a bestial four-legged starting point has obvious appeal to our ego. But beyond that, there's clearly something unique and significant about our two-leggedness. While there are other habitually bipedal animals—notably birds (and many of their extinct dinosaur kin) but also kangaroos, pangolins (scaly anteaters), and several rodents—we humans are nearly alone in our bolt-upright posture.* Furthermore, our acquisition of bipedalism has had particularly profound consequences, most obviously by freeing our grasping hands from the compromising demands of locomotion. This functional release allowed them to become the extraordinary manipulative organs that were to play such a major role in our rise to global dominance. That's not to say that the direct, locomotory benefits of the transition weren't equally important. As we saw in Chapter 1, our long-distance walking and sustained running skills are exceptional, particularly when regarded in the light of our superior thermal biology (itself a consequence of our bipedal posture). These abilities are heavily implicated in the development of our foraging and hunting skills, which in turn were probably instrumental in our social and cognitive evolution. Without a doubt, our transformation into full-time bipeds was one of the most significant transitions in the history of our lineage. If it hadn't happened, we wouldn't be here to discuss it. Small wonder *The March of Progress* strikes such a chord.

Given the great importance of our shift to two-leggedness, many evolutionary biologists are understandably keen to find out how and why it

* Penguins share this trait, but their tiny legs render their means of terrestrial progression vastly less effective than ours. Of course, the tables are emphatically turned underwater.

happened. Now, you might think that with the many beneficial consequences of bipedalism listed earlier, we already have the solution to this conundrum. If going around on two legs is so useful, surely the origin of human bipedalism was a straightforward task for natural selection to bring about? I'm afraid not. Such thinking grants foresight to the blind process of evolution. The future fitness consequences of an evolutionary transition, however beneficial, are completely irrelevant when the transition is actually occurring. If a trait is to spread in a population, its bearers must have more surviving offspring than the nonbearers. Nothing else will do. With this in mind, we can see that most of the convenient outcomes of hand-freeing (the selective drive favored by Charles Darwin in *The Descent of Man*) can only have become pertinent once we were already habitual bipeds, and so should be viewed more as a bonus than an impetus of the transition. Similarly, the ability to run down an antelope over many miles cannot have arisen until our quadrupedal ancestors were almost as ancient history as they are now. The first fruits of bipedalism were necessarily more subtle. If we are to find the vital selective nuggets that unlocked our forebears' path to humanity, we first need to examine how *incipient* bipedalism affects locomotory performance.

## WALKING BADLY—A STEP-BY-STEP GUIDE

Chimpanzees and bonobos, our closest living relatives, can walk on their hind legs, but they don't do it very often, for good reason. From tip to toe, they're simply not built to be bipedalists. First, their spine lacks the pronounced curvature of the lower back that's such a conspicuous feature of our vertebral column, thanks mainly to the great length of their hip bones (specifically the iliac blades, which are what you can feel when you put your hands on your hips). So, when walking on two legs, chimps can't position their center of mass directly over the hips, forcing them to keep the knees bent to avoid toppling forward. Their iliac blades also lie parallel to the back—they don't curl round the sides of the body as do ours. An important consequence is that chimps' gluteus medius muscles are hip extensors, whereas ours stabilize the pelvis on the supporting leg. Were they to walk like us, chimps would be unable to stop a sideways fall of the torso away from the stance leg; they instead adopt a lurching Trendelenburg gait to compensate. This problem is accentuated by chimps' bowed legs—they lack our valgus knee, and so can't tuck their legs right in under the body. Furthermore, their feet are permanently flexible, lacking the anatomical features (including the instep arch) that make our windlass mechanism possible. Not that such a stiffening mechanism would do them much good—their long toes would likely break if their tips were exposed to the high forces of a human-style toe-off, and the big toe is widely divergent. For these reasons, the propulsive push must come from the middle of the foot, not the toes.

To see how these features impact on locomotory efficiency, you might like to try walking like a chimp for a bit (privacy of your own home recommended). Bend your legs, keep them wide, tilt your spine forward, and go, remembering to relax your gluteus medius muscles and to keep your weight off the inside edge of your feet and big toes. It's surprising how physically (and mentally) demanding the simple act of walking becomes when we suppress our natural locomotory instincts. There are various reasons for this. First, there are too many movements that consume energy but don't contribute to forward progression (such as the side-to-side Trendelenburg swagger). Secondly, thanks to the semicrouched posture, the line of action of the ground reaction force lies far in front of the hip joint (Fig. 2-1). This is important, thanks to a simple home truth about levers. The force that you get out of a lever is not simply dependent on how much force you put in. As anyone who's ever opened a can of paint with a screwdriver can confirm, much depends on the relative lengths of the input and output lever arms—in other words, the distances between the lines of action of the input and output forces and the lever's pivot. If the input lever arm is, say, twenty times longer than the output lever arm (as may be the case in the paint can/screwdriver example), the output force will be twenty times larger than the input force. Technically, increasing the length of the input lever arm increases the *torque*: the force multiplied by the length of its lever arm. To return to our current investigation, when standing with a chimplike stoop, the lever arm of the ground reaction force about the hip is much longer than when standing upright, so its torque is larger, and the hip extensors, such as the hamstrings and the erector spinae muscles, have to work hard just to stop the body folding at the hips.*

The postural woes don't end there. Keeping the flexed knee stabilized means recruiting the bulky quads, adding yet more to the fuel bill. The potential leverage of our magnificent feet can't be fully exploited if the push comes

* On a different note, the main function of our kneecaps is to increase the torque of the quads: by diverting the path of the patellar tendon, the kneecap moves the line of action of the contraction force away from the hinge axis of the knee joint.

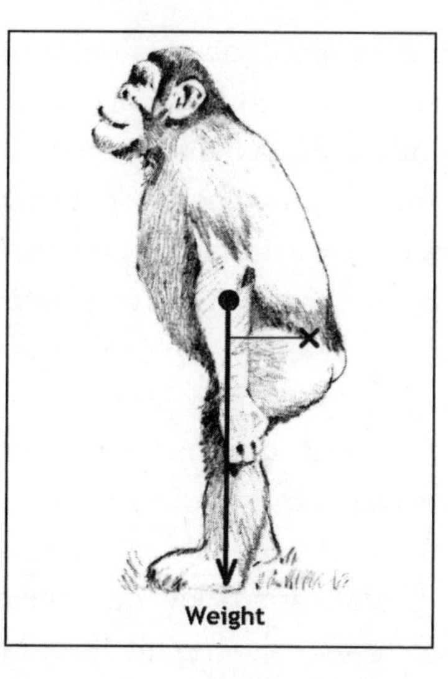


2-1: A chimpanzee's stiff, straight, lower back makes it difficult to stand up (or walk) straight—keeping the center of mass over the polygon of support requires a stooped posture. With the weight vector now situated quite far in front of the hip joint (x), the weight's torque about the hip is large, requiring considerable effort from the hip extensor muscles to stop the joint collapsing. In humans, the center of mass is much closer to the hip joint, reducing the costs of bipedalism.

from the midfoot rather than from the toes. Finally, walking with permanently bent knees and hips blocks our primary mechanism for energy conservation. Without the stiff pendulum-like motion of the legs, we don't get the usual efficient interchange of gravitational potential and kinetic energy. This all adds up to a substantially increased cost of locomotion. There is a flip side—the jarring heel-strike is eliminated—but this seems small recompense for all that wasted energy.

Bearing all this in mind, the results of a recent experiment run by Michael Sockol of the University of California, Davis, will come as no surprise. In his experiment, five chimps were trained to walk bipedally on a treadmill while wearing face masks to measure their oxygen consumption. Their average cost of locomotion, when corrected for body mass, was about 75 percent higher than for humans. What might be more surprising is that when Sockol tested the same chimps during their more usual quadrupedal gait, the energetic costs were nearly as high—only 10 percent lower on average—with two of five individuals doing better on two legs than on four. Any economic benefits connected with the increased stability of quadrupedalism seem to be nearly canceled out by the additional volume of muscle brought into play.

It doesn't help that in quadrupedal locomotion, the hindlimbs still carry 80 percent of the weight—due to chimps' long arms.

The chimp version of quadrupedalism is thus a far cry from the grace and economy of the feline or equine equivalent that we looked at briefly in Chapter 1. When it comes to explaining how the bipedal transition occurred, however, the inefficiency of chimps' gait is helpful to us, for it shows that some of the selective obstacles in the path of the change were less significant than we might suppose. The enormously improved efficiency of human locomotion implies that even minor adjustments to an initially costly gait could have paid dividends. Indeed, simple mathematical models have shown that a change in leg and muscle lengths of only 10 percent from a chimplike starting point would bring bipedal costs below quadrupedal costs. However, this immediately raises a troublesome question, a favorite of creationists: if this is the case, why are quadrupedal chimps still around?

This is an easy objection to bat away. First, locomotory economy, though important, is rarely the sole target of natural selection: speed or stability could have higher priority in the wild. Chimps can indeed keep up an enviable pace in dense undergrowth (though out in the open, a decent human runner would beat them hands down, or rather, hands up). Furthermore, chimps aren't exactly dedicated to the terrestrial life. While they do most of their traveling on the ground, they do most of their feeding in trees, and because the locomotory challenges of the arboreal realm are particularly acute for a creature the size of a chimp (a single misstep could be disastrous), it is the arboreal side of their locomotory repertoire, not the terrestrial, that dominates their mechanical design. Their long, powerful arms and hands, quite apart from extending their reach, allow them to hold on to and climb tree trunks with minimum muscular effort. That's because long arms can be oriented more vertically than short arms when holding a tree trunk, and so will tend to bear more weight in tension than in bending. That tension can be partially resisted passively by joint ligaments, whereas an arm subjected to bending loads must be resisted actively by the muscles. Try supporting your body off the ground with outstretched vertical arms, then with outstretched horizontal arms, and you'll feel the difference. Looking at the other end of the body, chimps' permanently flexible feet, with opposable big toes and long, curved, lateral toes, obviously come into their own in the trees, by providing a secure grip on trunks and branches. Their long, flat hip bones and concordantly long hip muscles allow them to extend their thighs powerfully from the fully flexed configuration—an activity for which our shallow, curved pelvis is ill-suited. Chimps are therefore much better than us at squat thrusts, and more pertinently, they're much better at climbing up tree trunks.

## HOW IT MIGHT HAVE HAPPENED

Tweaking any of the chimp features in a human direction to improve locomotion on the ground would inevitably worsen performance in the trees. But take the trees away, and this negative side effect becomes irrelevant. According to the classical explanation of the evolution of human bipedalism, that is exactly what happened. The Miocene and Pliocene epochs (from 23 to 2.5 million years ago) were marked by a prolonged global cooling and drying, and saw the spread of grasslands at the expense of forests throughout Africa (our ancestral home at the time). Our last common ancestor with chimps, so the story goes, was thus faced with a "choice": stick with the shrinking forests or head out into the savannah. The chimp line chose the former route and stayed as they were; the human line—the hominins*—chose the latter and were transformed into efficient terrestrial bipeds.

We now encounter our first conundrum: Why should going terrestrial entail going bipedal? Why not refine the existing quadrupedal mode instead? This question is given added force by the observation that the apes' closest living relatives—the Old World monkeys—include many species (most notably baboons) that are committed quadrupedal terrestrialists, with features strongly reminiscent of typical terrestrial mammals: they have long, slender legs, relatively rigid feet, and permanently stand and walk on their toes and fingers. Indeed, terrestrial features are so prevalent in the group that it's now widely accepted that the last common ancestor of all living Old World monkeys was a ground dweller.

Such objections have led some to question whether the bipedal transition was driven by locomotory considerations at all. After all, there are other savannah-specific benefits. For one, standing upright gives you a better view of the distant surroundings. On a related note, being bipedal may enable more effective sexual displays, providing a potential explanation for a couple of unusual human characteristics: permanently swollen female breasts and the unusually large male penis. Unfortunately, while the adaptive logic is

* *Hominin* is the preferred term for everyone on the human side of the human/chimp divergence. The more familiar term *hominid* is now defined more broadly, to include all the great apes.

sound, both theories suffer from the likely weakness of the necessary selective pressures. Even meerkats, which are famous for their bipedal lookout behavior, move around quadrupedally.

A more promising idea is that bipedalism evolved for thermal reasons. On the open savannah, any animal runs the risk of overheating in the intense tropical sun, but this risk can be ameliorated by reducing the area of the body exposed to direct sunlight. An upright stance and gait achieve just this, while also exposing the body to stronger cooling air currents (wind speed is always lower near the ground due to friction between the ground and the air). This ties in rather nicely with our hairlessness and sweatiness, and with our later persistence hunting technique. One could always raise the objection that if these thermal considerations are so important, other savannah inhabitants should have gone bipedal as well. However, given an initial bipedal propensity (remember the costliness of quadrupedal locomotion in chimps), such pressures could have been sufficient to push hominins over the edge, so to speak, at which point locomotion would have taken the selective driving seat. To paraphrase Darwin, it is at least a theory by which to work. So far, however, we've looked only at living species. The time has come to consult the fossil record to see what our extinct relatives have to say on the matter.

## I LOVE LUCY

Unfortunately, the human fossil record leaves much to be desired: specimens are few and far between, and most of those are fragmentary. However, every so often a more complete skeleton is unearthed that shines a brighter light on the distant hominin past. The most famous of these was found in 1974 in the Afar region of Ethiopia, near the village of Hadar in the Great Rift Valley, by Donald Johanson, then curator at the Cleveland Museum of Natural History. He was part of a multinational team that had been assembled to sweep the area for clues of hominin presence. The oft-told story is a classic tale of fortune favoring the prepared mind. On the morning of November 24, Johanson decided on a whim to recheck the bottom of a small gully, and was about to leave when he caught sight of a broken piece of bone lying on the slope. A brief exploration of the area revealed more fragments, so in the afternoon he returned with the entire expedition team, who managed to recover several hundred pieces over the following three weeks. Amazingly, all came from the same skeleton—a 3.2-million-year-old, 3.6-foot-tall female (deduced from the width of the pelvis), belonging to the species

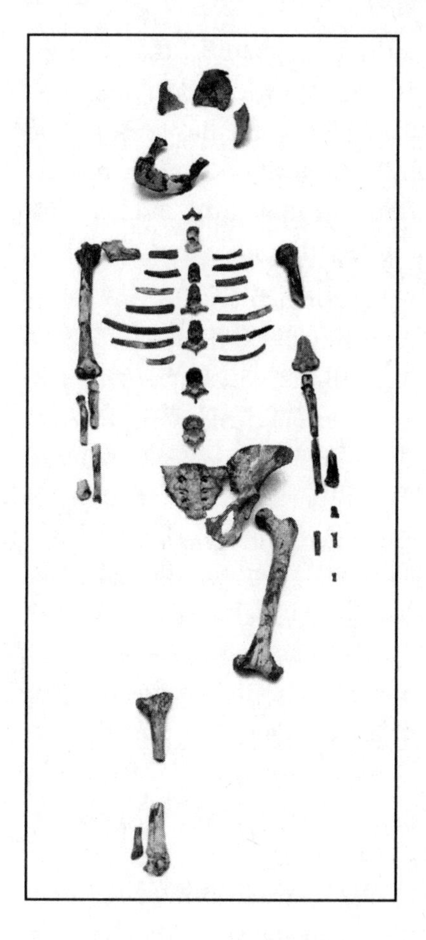

2-2: Lucy—the most famous example of *Australopithecus afarensis.*

*Australopithecus afarensis* (southern ape from Afar), a possible human ancestor.* Officially labelled specimen AL 288–1, the fossil is more widely known by her nickname—Lucy—which she acquired courtesy of the Beatles hit "Lucy in the Sky with Diamonds" that was playing on the camp radio during the evening's celebrations.

Lucy was at the time of her discovery the oldest hominin known, and with 40 percent of her skeleton recovered, she had enormous potential to, dare I say, elucidate the evolution of human bipedalism. Early indications

* I'm usually highly suspicious of claims that fossils are direct ancestors of living species (the "missing link found" stories so beloved of the press), as they implicitly assume a low likelihood that lineages go extinct. The fossil record emphatically contradicts this assumption—it's a panoply of evolutionary dead ends. The recent claim that "Ida," a 55-million-year-old primate, was a direct ancestor of humans was ludicrous hype. However, the younger the fossil species, the greater the likelihood that its descendants are still with us, and 3.2 million years seems recent enough to at least entertain the possibility of ancestral status for *Australopithecus afarensis.*

were that she was a habitual biped, with a gait more humanlike than chimplike, for she had our valgus knee. Her pelvis was also wide and shallow—a far cry from the tall iliac blades of apes—so her gluteus medius muscles had already shifted into a humanlike pelvis-stabilizing position. But there were inconsistencies. Her legs were quite short by human standards, and her arms rather long—roughly halfway between chimp length and human length in relative terms—with long, curved finger bones. This was curious, because such arms seem well suited to an arboreal existence, and climbing arms on a habitual biped were unexpected: the shift to upright bipedalism was supposed to have been driven by an exodus from the forests. There are three possible explanations for the odd juxtaposition: (1) the functional interpretation of Lucy's arms may be wrong—perhaps we're just seeing a midtrend snapshot of their gradual shortening in the hominin lineage; (2) Lucy wasn't a proficient biped, but a more chimplike jack-of-all-trades, still hanging out in forests; (3) if neither of those suggestions are supported, the whole forest-to-savannah explanation for the origin of bipedalism will need some serious revision.

Sadly, Lucy is missing her feet, depriving us of a vital diagnostic part of her locomotory kit. However, in 1976—the very same year that she was introduced to the modern world—another discovery was made, largely by accident, that more than compensated for this shortcoming. Two paleoanthropologists on a fossil-finding expedition in Laetoli, Tanzania, were having an elephant dung fight (recreational facilities presumably being rather limited) on the way back to camp. One of the combatants, while searching for ammunition, came across a series of depressions in a hardened ash layer, which on closer examination turned out to be animal footprints. The find didn't elicit much excitement at the time, but the following year, a more thorough search of the area yielded several preserved animal trackways, some of which looked remarkably like human footprints. That alone would be nothing to write home about, but a combination of radiometric dating and the identification of mammalian fossils in the ash bed indicated that the tracks were over 3.6 million years old.

There's something deeply moving about fossilized prints. A few years back I was fortunate enough to visit a similar, though much older find in Crayssac, in the south of France, where dozens of pterosaur tracks had been revealed crisscrossing a Jurassic beach. It's difficult to describe how it felt to walk across the very same ground that the ancient creatures had dinted so long ago. When those long-dead individuals belong to a species possibly ancestral to ours, the experience must be truly magical. The fact that the

Laetoli prints comprise two trackways running alongside each other only adds to the poignancy of the find, as did the realization that a third individual had been following the first, marking its footsteps like Good King Wenceslas's page. It's impossible not to envisage a small family, maybe on a journey to a waterhole, with mom and dad walking hand in hand while their child struggles to keep up behind, though of course we'll never be able to tell whether the three individuals were related. We can, however, have a reasonable stab at identifying what species they belonged to. The timing alone makes *Australopithecus afarensis* the most likely candidate of the known hominins, a hypothesis bolstered by the close similarity between the Hadar material and fragmentary skeletal remains from the Laetoli site. Without a fossil of the printmaker lying at the end of the tracks we can't be sure, but for now it seems fairly safe to conclude that Lucy and the Laetoli ramblers belonged to the same species.

The Laetoli prints are strikingly modern in appearance—more or less indistinguishable from human prints to a nonexpert. A closer look reveals that the big toe was at a slight angle to the axis of the foot, unlike ours, although there's no indication of a widely divergent opposable big toe as seen in apes. But it would take a bit of twenty-first-century technology to wring all the information from the impressions. David Raichlen of the University of Arizona and his colleagues used a 3-D laser scanner to construct detailed virtual casts of the prints, and then asked volunteers to walk barefoot in specially prepared wet, fine sand—a reasonable proxy for the ancient damp ashfall at Laetoli—before scanning their footprints, too. Critically, the volunteers were instructed to use two different gaits: their normal erect, stiff-legged walk and a more chimplike bent-hip, bent-knee shamble. The differences between these modes left telltale signs in the sand. With bent legs, there was a prolonged application of force by the forefoot, and a consequently deeper impression under the toes. With straight legs, the heel and toe impressions were roughly equal in depth or the heel slightly deeper. The differences were small but significant and were always detectable even as the water content of the sand was adjusted. The Laetoli tracks unambiguously matched the erect-gait prints. These results were supported by a number of virtual reconstructions of Lucy's walking style,* which showed that her

* Chief among these were several collaborations between Robin Crompton at the University of Liverpool, Bill Sellers at the University of Manchester, and Weijie Wang at the University of Dundee (and others), and an independent study carried out by Akinori Nagano and his colleagues at Arizona State University.

musculoskeletal system permitted an efficient erect gait. There can therefore be little doubt that *Australopithecus afarensis* walked rather like us, a conclusion later confirmed by the long-anticipated discovery of some fragmentary foot bones, which indicated the presence of a humanlike instep arch.

The supposed arboreal side of Lucy and co.'s existence is more difficult to confirm. Of course, tree-climbing leaves no prints, and arboreal locomotion is too complex and varied to be usefully modeled on a computer. Therefore, we're left with the traditional method of finding biomechanical clues in the fossils. Unfortunately, we have little to go on beyond the aforementioned long arms and curved fingers, other than the simple inference that the downgraded opposability of the big toe and shortened pelvis meant that Lucy was not as accomplished a climber as chimps. What we do know is that australopithecine morphology remained fairly stable for about 2 million years—up to the appearance of our *Homo* genus—which argues against the midtransformation snapshot theory for the long arms: they must have had some use to have been retained for so long. One suggestion is that *Australopithecus*'s climbing activities were restricted to simple nighttime trips into the canopy to find a safe place to sleep.

Of course, any climbing skill is of little use without something to climb, so some indication of trees in the neighborhood would be a useful addition to the body of evidence. While we don't have any fossilized trees, fossil mammals of the right age at Hadar, Laetoli, and other *Australopithecus afarensis* sites indicate medium-density woodland habitats, with a gradual shift to more open wooded grasslands later on. Regardless of *Australopithecus*'s arboreal facility, we are therefore hit with a bombshell: the habitual bipedalism of *Australopithecus afarensis*, and by extension the whole hominin lineage, must have evolved in a woodland context, not savannah. It was the subsequent refinement leading to the appearance of the ultraefficient *Homo ergaster* grade that was associated with savannah habitats. Our original scenario is way off the mark.

There's been no shortage of attempts to account for the apparent evolution of terrestrial bipedalism in a forested setting. One suggestion, recently given life by Kevin Hunt of Indiana University, was that it was initially a feeding adaptation. Hunt noticed that chimpanzees and bonobos often stand upright to reach for overhead fruit, and proposed that if this activity were to become habitual, natural selection might cause a shift to an australopithecine morphology. Special attention was drawn to Lucy's shallow pelvis, which was interpreted as an adaptation for standing rather than walking. The theory makes some sense, but fails to account for some conspicuous australopithecine features. Why, for instance, would selection just for standing have caused our ancestors to get rid of their extremely useful opposable big toes, when, if anything, they'd have been better off keeping them for such purposes?

Another intriguing suggestion, again based on observations of chimps, was recently made by Carsten Niemitz of the Institute for Human Biology and Anthropology in Berlin. He proposed that habitual bipedalism was a wading adaptation, and has amassed an impressive body of supporting evidence. First, chimps and bonobos often wade bipedally. Then there's the circumstantial evidence that almost half of today's primate species are at least occasional shore dwellers, along with paleoecological indications that lakes and rivers were present in the localities of many of the early hominins, Lucy included. Tropical freshwaters generally provide plentiful high-quality food year-round, giving an impetus for a behavioral shift away from an ancestral fruit-eating lifestyle that was perhaps becoming less tenable as the forests thinned. From a mechanical perspective, walking bipedally in water, though costly and slow, is forgiving of a misstep and provides support for submerged body parts.

There are obvious shades here of the infamous aquatic ape hypothesis recently championed by Elaine Morgan, although Niemitz's ideas rest on surer scientific foundations: he considered all the evidence available, whereas Morgan ignored a great swathe of information that contradicted her central premise. That said, his theory is not without its problems. For instance, he suggests that selection for longer legs in a wading context, either for locomotory or visual reasons (a higher point of view makes it easier to pinpoint submerged objects), eventually compromised *terrestrial* quadrupedal locomotion, thanks to a fore-/hindlimb mismatch. This idea is difficult to reconcile with the fact that all our ape relatives have longer arms than legs: a relative increase in the length of the latter should have made quadrupedal locomotion easier, not more difficult. Still, such objections aren't fatal to the idea that semiaquatic activities were important in our locomotory evolution—indeed, this might have proved a winning theory, were it not for the discovery of an extraordinary fossil that would conclusively blow any notions of semiaquatic origins of bipedalism, well, out of the water.

## THE NEW OLD KID ON THE BLOCK

At 3.2 million years old, Lucy lies roughly halfway between the likely date of our divergence from chimps and the present day. From her time onward, our lineage's history is relatively well known, with about twenty recognized species, mostly in the genera *Australopithecus* and *Homo*. The same cannot be said of the earlier half of hominin evolution: just a handful of specimens have been found, representing four or five species, and until recently, all were known only from bits and pieces. The advanced state of Lucy's bipedalism made this state of affairs doubly frustrating, for it means that the all-important transition to an upright gait was nearly complete before 3.2 million years ago, during the dark ages of the hominin fossil record. Many of the early hominin fragments bear tantalizing hints of habitual bipedalism, though all have been contested. For instance, CT scans of the femur of the 6-million-year-old Kenyan species *Orrorin tugenensis* show a thickening of the bone on the underside of the femoral neck (the bony bridge between the spherical head and the shaft), suggesting that the leg habitually bore a relatively high proportion of body weight, which is consistent with a bipedal gait, but the validity of the scans has been disputed.

Such ambiguous clues were never going to resolve the origins of bipedalism. What we dearly needed was another Lucy—a partial skeleton. Lost cause though this may have seemed, the only thing the paleoanthropologists could do was keep looking. In 1992, against all the odds, they struck gold. As usual, it all began when a small fragment was spotted by an eagle-eyed fossil hunter. The fragment in question was a molar, found near the village of Aramis, just 46 miles from Lucy's last resting place, by Gen Suwa, a member of a team led by Tim D. White of the University of California, Berkeley. Suwa quickly realized that the tooth was hominin, and a protracted search of the area was initiated to find as much of the owner as possible. Thanks to the extraordinary thoroughness of the field crew (collaborator Owen Lovejoy from Kent State University commented that they seemed to "suck fossils out of the earth"), a respectable 45 percent of the skeleton was recovered, including most of the hand and foot bones, making it even more complete than Lucy. The fossil (female again, as it happens) was named *Ardipithecus ramidus*—Ardi for short—incorporating the local words for ground (*ardi*) and root (*ramid*), in deference to her basal position in the hominin family tree. Although this was an unbelievably lucky find, there was a catch. The fossil was desperately fragile—it literally crumbled to the touch, and some parts of the skeleton had fragmented into over a hundred tiny pieces. Even when extracted—itself a mammoth task requiring endless applications of consolidating adhesive and the bulk removal of entire blocks of sediment—it took nearly fifteen years of painstaking microscopic preparation and state-of-the-art computer-assisted reconstruction, chiefly by Owen Lovejoy and his team, to put Ardi back together again. In 2009 she was finally unveiled to the expectant scientific community in a special issue of the journal *Science*.

First things first—Ardi was estimated to be about 4.4 million years old, so she was around fairly soon after the chimp/human divergence. One would thus expect her to have been more chimplike than the later hominins like Lucy. In many respects, she met these expectations: she had a chimplike protruding snout, arms that were slightly longer than Lucy's in relative terms, and, alone among hominins for which the foot is known, she had a divergent, grasping big toe. But other aspects of her anatomy told a somewhat different story. Like humans, she had six lumbar vertebrae, whereas living apes have only three or four, indicating a rather mobile back by ape standards. While her hands were expansive, they were much more humanlike in their proportions than are chimp hands. Notably, her metacarpals (hand bones) were even shorter in comparison with her fingers than ours—not at all like the elongate metacarpals of chimps—and her thumbs were much more robust than the spindly, stunted ape equivalents. Even Ardi's feet, despite that grasping big toe, were quite different from those of apes, being rather stiff, if not quite as stiff as ours. Her pelvis presented even more of an enigma. From the hip joint downward it resembled the pelvis of a chimp, but the rest was hominin through and through, with short, laterally facing iliac blades.

Ardi's chimeric hips sum up her entire anatomy. Her long arms, opposable big toes, and chimplike lower pelvis reveal that she was an adept climber—much more comfortable in the trees than was Lucy—but her more humanlike iliac blades tell us that her gluteus medius muscles were, like ours, pelvis stabilizers, so she could have walked bipedally much better than a chimp can. This conclusion is borne out by her relatively stiff feet, which

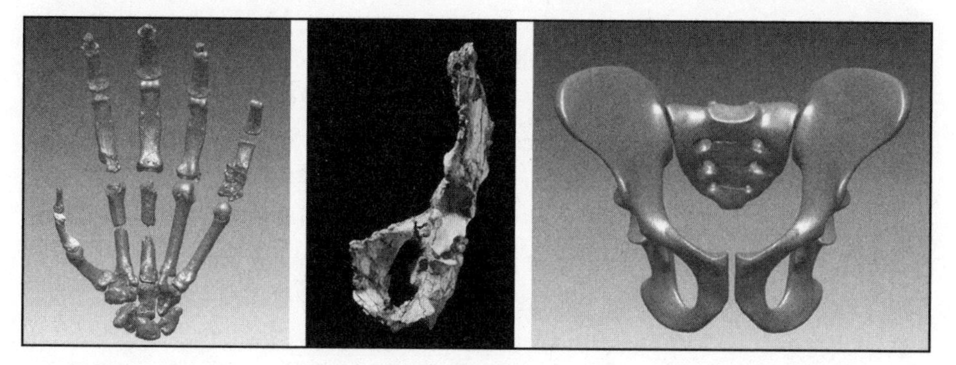

**2-3:** The fossilized hand (left) and pelvis (center), and the reconstructed pelvis (right) of Ardi—*Ardipithecus ramidus.*

doubtless functioned reasonably well as propulsive levers (although, without the assistance of the big toe, she would never have beaten Lucy in a race on the ground). Most significantly, her flexible lower back, unconstrained by her short iliac blades, would have allowed her to curve her spine to bring the center of mass directly over the hips when on two legs, so she needn't have kept her legs bent or back inclined while walking.

## TWO LEGS IN THE TREES

From a certain point of view, Ardi's mishmash of characteristics is nothing out of the ordinary. Treating her as an ape/human intermediate, we might expect some aspects of her anatomy to be more humanoid, others to be more apish. If, however, we temporarily ignore her intermediate status and simply take her as she is, we arrive at a somewhat different conclusion. All the evidence indicates that Ardi managed to combine effective bipedal *and* arboreal locomotory modes, which is a somewhat surprising inference, given what we know about chimps. But before we get too flummoxed, we might want to take a break from chimps for a moment and take a glance at our Southeast Asian ape relatives: the gibbons and orangutans. Gibbons, you see, though principally known for their method of swinging acrobatically from arm to arm (brachiation) at breakneck speed through rain forest canopies, are also the most bipedal of living apes, and readily adopt this mode to walk or run along branches or across open ground.

Orangutans are even more enlightening. They, too, are dedicated tree dwellers, with a complex, adaptable approach to canopy locomotion: they use a wide variety of modes ranging from upright clambering to slothlike four-limbed suspension. But they can also move bipedally, and uniquely among the living apes, they usually do so with erect, straight legs, which enables them to exploit the pendulum-like interconversion of gravitational potential and kinetic energy, as do humans. They don't do it very often—in only about 7 percent of locomotory bouts, according to a recent Sumatran field study by Susannah Thorpe of the University of Birmingham—but it's their preferred mode for moving on thin, flexible branches, a particularly critical aspect of their environment that must be negotiated to access most fruit (their favorite food) or cross from tree to tree at canopy level. Bipedal locomotion has an advantage over quadrupedal in this context, in that it allows the weight of the body to be supported while at least one arm is free to extend toward a juicy fruit or swaying branch.

No one is suggesting that Ardi was essentially a 4.4-million-year-old orangutan, or that the Asian great apes have remained unchanged since the time of our last common ancestor. Orangutans obviously have many specializations, such as their elongated arms, that contribute to their arboreal locomotory flexibility. But they show us that there's no inherent conflict between fairly efficient bipedalism and a life in the trees. Indeed, as pointed out by Robin Crompton—a human and primate evolution expert at the University of Liverpool—a somewhat bipedal arboreal lifestyle is probably the default mode for apes. That's because, human or ape, we all tend to orient the spine vertically rather than horizontally: in technical terms, we're habitually orthograde (from the Greek *orthos*, meaning "upright") as opposed to pronograde (from the Latin *pronus*, meaning "bent forward"). We humans, of course, exhibit this posture when we walk or run; apes do it whenever they're suspended from their arms or climbing vertically. Monkeys aren't like this—they're typically pronograde when on the move, scampering along branches and suchlike.

What was it that drove the apes to adopt a more upright posture than their monkey ancestors? There's a clue in the fossil record. The last common ancestor of the living apes probably lived about 17 million years ago, based on the genetic distinctness of its modern descendants. While we're never likely to find this creature, the recently discovered 13-million-year-old *Pierolapithecus* from Spain is a reasonable proxy. This specimen (a male, for once) seems on the basis of his wide, shallow rib cage and certain aspects of his vertebrae to have had an orthograde posture, as we would now predict. Significantly, he was also a giant by monkey standards, with an estimated mass just shy of 67 pounds. Monkey-style above-branch quadrupedalism would have been quite unsafe for such a bulky animal, for the polygon of support—the area bounded by the supporting limbs—would have been precariously narrow. This is why no arboreal monkey reaches this sort of size. By going orthograde and bringing the center of mass over the feet, the early apes were free to spread their arms wide when clambering about the treetops, considerably reducing the risk of a fall, and making a bit of bipedalism all but inevitable.

But that's not all. Our old ape friend has hidden in his skeleton more clues than his size and posture alone, but to interpret his secrets we're first going to have to get to grips with, well, gripping. *Pierolapithecus*'s hands, you see, were rather like Ardi's in their overall proportions. Such appendages are perfectly adequate if you tend to support yourself from below. But one of the great things about orthogrady is that it allows you to support your weight from above—in suspension. Hooklike hands with elongated metacarpals and curved finger bones are ideal for this sort of thing, as are extra-long arms. Big thumbs, however, just get in the way, so the tinier the better. This suspensory anatomical scheme has been developed to various degrees in all modern apes, from the more conservative gorillas to the extremist gibbons. But another, subtler evolutionary tweak ensures efficient suspension. Maintaining the hooklike posture of the hand will be much easier if the finger flexor muscles are short. That's because while the force with which a muscle can pull depends on its cross-sectional area, its energy expenditure while pulling depends on its volume and therefore its length; shortening the finger flexors thus allows an ape to dangle from a branch more economically and with less chance of muscle fatigue. The flip side is that fingers with short flexors can't be bent back at the knuckles, making the old palm-down style of quadrupedalism seen in monkeys impossible in suspensory specialists. Therefore, should a suspensory ape wish to go four-legged on the ground for a bit, it must use some other hand-contact scheme, such as walking on the knuckles, which is what chimps, gorillas, and orangutans all do in this situation (gibbons never go quadrupedal—their arms are just too long).

The fact that suspensory adaptations are seen in all modern apes instantly implies that their last common ancestor was similarly a dangling enthusiast. However, the conservative hands of *Pierolapithecus* and Ardi tell a very different story. Not only are their metacarpals a little on the short side for habitual dangling, but the disposition of the joint surfaces of their knuckles indicates that both creatures were able to bend their fingers back, like monkeys, and indeed us, and so must have had relatively long finger flexor muscles. With such hands—more suited to pushing up from below than pulling up from above—they would have found prolonged bouts of suspension difficult, but they would also have never needed to knuckle-walk. This inference is supported by the lack of any trace in these creatures of the bony buttressing that's present in the hands of knuckle-walkers to help the fingers and wrists cope with the unusual weight-bearing regime. Now, one could argue that what we see in these two extinct apes is the result of an evolutionary reversion—a harking back to a presuspensory past. But *Pierolapithecus* and Ardi are old for their lineages—they were around soon after the last common ancestors of, respectively, the modern ape + hominin group and the chimp + hominin subgroup. Unless evolution did a very rapid whitewashing job, those ancestors must also have had conservative hands that weren't specialized for suspension as in the modern apes, but were instead used for more generalized clambering.

It would thus appear that suspensory adaptations evolved independently in the gibbon, orangutan, gorilla, and chimpanzee lines. Surprising as this may seem, it makes sense of the fact that each of the modern apes has a rather different take on a suspensory way of life. Gibbons went for a decrease in size and brachiation. Orangutans plumped for arboreal locomotory flexibility. Chimps and gorillas, on the other hand, emphasized vertical climbing, and independently modified their hips in a way that now enables powerful leg extension but also precludes erect bipedal walking. The African apes combined these changes with an increased exploitation of the terrestrial environment for tree-to-tree travel (which was doubtless why the ability to climb vertically up tree trunks became so useful), and so became accomplished knuckle-walkers.

Now we're finally in a position to appreciate Ardi's true significance. Simply put, she tells us that what made hominins special—the attribute that eventually led to their habitual terrestrial bipedalism—wasn't the adoption of two-leggedness itself, for that had effectively already happened in the trees long before, when the early apes went orthograde.* Rather, it was the hominins' refusal to jump wholeheartedly onto the suspensory bandwagon; instead, they emphasized the bipedal aspect of the ancestral locomotory suite. Following Ardi's divergence, they gradually downgraded their arboreal abilities and continued to push for greater bipedal efficiency—shortening and reorienting the hips, angling the thighs toward the midline, drawing in the big toes, beginning to arch the feet, and developing a permanent curvature of the lower spine. Lucy illustrates this grade. Later, the arms were shortened, the legs lengthened, the Achilles tendon—fairly insignificant in apes—was elongated and thickened, and the windlass mechanism associated with the instep arch was refined. Thus did the honed two-gear terrestrial bipedalism of *Homo ergaster* and its descendants appear.

Why the various anatomical experiments of the different ape lineages occurred at their particular times and in their particular places isn't known for sure, although the gradual opening up of the African forests in the Miocene doubtless played a role in the increased terrestriality of the African apes and hominins. The gibbon solution, on the other hand, may well have been a uniquely apish response to certain opportunities available in Southeast Asia (you'll learn about those in Chapter 3). Another intriguing and by no means mutually exclusive possibility is that the impetus for the various evolutionary

* If we accept that upright bipedalism evolved long before the chimp/human divergence, we have no reason to dispute the bipedal status of the early hominins, such as *Orrorin*.

changes in the apes came from the Old World monkeys. These, you'll remember, went terrestrial early on—shortly after the two groups diverged about 30 million years ago—supposedly leaving the arboreal realm to their ape cousins. Later, however, a few of them apparently had second thoughts and reinvaded the trees. By all accounts they seem to have nearly driven the apes to extinction, maybe because they were smaller (with higher reproductive rates), or had less stringent dietary requirements. Ape diversity is now a mere fraction of what it was in the Miocene, whereas the Old World monkeys are doing better than ever. According to this scenario, the apes that remain are extremists, displaced into fringe niches by the monkey onslaught.

Whether or not this story of betrayal, innovation, and plucky resilience is true (the apes may have been whittled down by other unknown factors, and the vacated niches occupied by monkeys later), what's now clear is that the *March of Progress*, for all its iconic appeal, got it completely wrong. The last common ancestor of chimps and humans wasn't very chimplike at all, as Darwin, to his endless credit, warned us might be the case in *The Descent of Man*. If anything, the remaining apes were the real innovators. Hominins merely refined something their ancestors had been able to do for at least 10 million years—ever since their increasing size caused them to literally turn their back on monkeyish scampering.

## THUMBS UP

Of course, much of this argument should be regarded as provisional—Ardi shows us that it takes but one fossil to pull the carpet out from under our feet. For the time being, however, it would appear that we have much to celebrate. Except that we've really only scratched the surface of an explanation of how we humans came to be. So far, we've uncovered the two coupled events that were instrumental in setting our primate ancestors on the road to habitual bipedalism: the size increase that occurred in the early apes and the consequent shift to orthograde clambering. But that door would never have opened had the primates lacked grasping hands and feet. Opposable thumbs—though often casually held up as key human accoutrements—are borne by most primates, including our most distant prosimian cousins, such as lemurs, so they must have evolved right at the base of the primate adaptive radiation. Given the critical importance of this trait in enabling our appearance, I feel we should hold the champagne until we've at least tried to explain its origins.

At first glance, this might seem an easy prospect. We primates are a quintessentially arboreal group, and for many years our hallmark grasping hands and feet were considered to be such obvious adaptations for safe navigation in the trees that little more needed to be said. The only problem with this idea, as originally pointed out by Matt Cartmill of Boston University in the 1970s, is that plenty of nonprimate arborealists lack strongly opposable digits. These include squirrels and colugos (close relatives of primates also known as flying lemurs, though they're not lemurs and, strictly speaking, they glide rather than fly). Instead of relying on a strong grip to avoid a slip or fall, they have crampons—their claws. Claws are truly ancient—they date back at least to the earliest reptiles that lived about 320 million years ago, and quite possibly evolved long before. Why did we primates get rid of our old grappling hooks, replace them with flat nails, and reinvent the wheel with a new nonslip system?

For once, we don't have to rely solely on the primates to help us answer this question. We've been blessed with an array of living nonprimates that have converged to some extent on the grasping primate solution, and so give us some useful comparative evidence as to its adaptive significance. They include tree frogs, some lizards such as chameleons (with their fused "double-thumbs"), and a number of marsupials, namely, the woolly opossum of South America and some of the possums of Australasia. If we can find some aspect of locomotory ecology that these groups share to the exclusion of nongraspers, and if we can make mechanical sense of the correlation, we will have gone a long way toward explaining why primate opposable thumbs and big toes evolved.

The first bit of ecological evidence we can glean is that all the nonprimate graspers are arboreal—no surprises there. But there's another common theme in their ecology that's shared with arboreal primates but not with squirrels and the like, and that's a preference for moving around on relatively fine branches. Grasping adaptations make complete sense in this context: the smaller the branch, the more likely it is that with appropriately adjusted joint surfaces, you'll be able to get a hand or foot around it. Any creature that can gets a number of important benefits. Chiefly, while claws work well on wide surfaces, they do little to prevent a fall while walking on narrow branches, when a creature's center of mass could easily stray to the left or right of its polygon of support. The friction of an encircling hand or foot can counteract such tendencies with comparative ease. This is of no small importance—the ability to walk safely along narrow branches grants free access to the periphery of tree crowns, where most of the fruit is to be found, as well as flower-borne nectar and the insects that feed on it.

The usefulness of grasping for fine-branch locomotion was recently verified in a neat if slightly mean experiment carried out by Pierre Lemelin of the University of Alberta and Daniel Schmitt of Duke University. They compared the performance of woolly opossums (graspers) and closely related short-tailed opossums (nongrasping inhabitants of forest floors) in a lab-based arboreal obstacle course. To successfully complete the course, the opossums had to cross from one tree to another, and as all individuals were released into the canopy, the most direct way of doing so involved negotiating a gap of roughly 12 inches between fine branches. The woolly opossums bridged the gap with ease, and had no trouble exploring the periphery of the canopy of the second tree, where tasty morsels of fruit had been left by the experimenters. It was a different story for their nongrasping relatives. All four individuals tested fell at least once, they usually had to adopt a desperate-looking clinging posture on fine branches, with arms crossed around the support, and more often than not they would head straight to the trunk to take the longer route across the floor.

The parallels between the marsupial graspers and the primates are striking. However, there are some important differences. For one, only the big toes have nails in the marsupials—all the other digits bear their original claws. More troublesome is the fact that these animals generally don't have fully opposable thumbs—opposable big toes seem to be sufficient. Intriguingly, this half-hearted embracing of a grasping way of life is shared with a couple of groups of close primate relatives: the tree shrews* (which aren't actually shrews—they just look a bit like them) and the extinct plesiadapiforms, which had their heyday 65 to 55 million years ago. The marsupial graspers might therefore be giving us a glimpse of the preprimate way of life, making them even more significant than they at first appeared.

The fact that woolly opossums and the like can deal with the arboreal fine-branch niche perfectly well without the full set of primate hand/foot adaptations tells us we're missing something: there must have been an additional selective push that caused the protoprimates to convert all of their claws to nails and develop full thumb opposability. Christophe Soligo of the

* Malaysian pen-tailed tree shrews have the dubious honor of being one of the few wild mammal species that habitually consumes large quantities of alcohol, in the form of fermented nectar from bertam palms (3.8 percent proof).

Natural History Museum, London, and Robert Martin of the Field Museum, Chicago, may have uncovered the critical extra selective factor. They undertook a careful analysis of the family tree of primates and our closest relatives, and identified a trend toward increased mass in our ancestral lineage, culminating in an estimated 2.5 pounds for the last common ancestor of all living primates (albeit with a wide margin of uncertainty, from 0.5 to 8 pounds). As we saw with the apes, size matters, but in this case it's because the bigger you are, the less likely it is that you'll encounter ungraspably large branches, and the more critical grasping skills become; adding prehensile hands to the existing prehensile feet is exactly what you'd predict natural selection to do in this situation. The same selective forces would be expected to widen the tips of the fingers and toes to increase friction. That's important because a wide fingertip bone can't support a typical claw. Nails may therefore be nothing more than a side effect of the evolution of an improved grip.

If Soligo and Martin are right, small primates, such as the prosimian nocturnal, moon-eyed bush babies and tarsiers, must have evolved from larger ancestors. That might explain tarsiers' unusually long fingers and curious crooked-fingered grip, as well as bush babies' delightful habit of urinating on their own hands and feet (something to consider if you're ever tempted to keep one as a pet)—all may be friction-increasing strategies to allow these miniature primates, whose larger ancestors had converted their claws to nails, to cope with relatively wide supports.† By the way, while I often encourage you to try out the locomotory strategies of other animals, I emphatically do not recommend peeing on your own hands and trying to climb a tree!

One further parallel between the marsupial graspers and primates is worth mentioning, although this one is so subtle that you'd almost certainly miss it if you weren't looking for it. While the prize for outlandish primate locomotion must go to the bipeds, even the quadrupeds among us have a slightly unusual way of walking. Most mammals walk with what's called a lateral sequence gait, in which the left/right alternation of the forelimbs is slightly delayed with respect to that of the hindlimbs: the sequence of footfalls is right hind, right fore, left hind, left fore, and so forth. It's a wonderfully stable sequence, because a large triangular polygon of support can be

---

†A similar challenge faced the tiny marmosets and tamarins of the New World, which, unlike tarsiers and bush babies, have taken the more obvious route of reevolving claws on all digits except the big toe. This might also have something to do with their unusual diet (the sap that exudes from damaged tree trunks), which gives them cause to hang out on relatively broad, vertical supports.

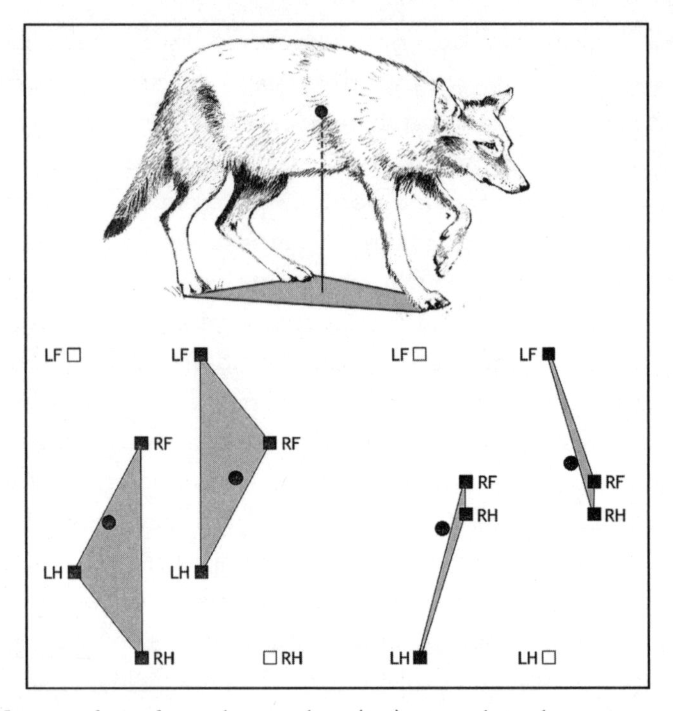


**2-4:** Most quadrupeds, such as wolves (top), use a lateral sequence walking gait: left hind (LH), left fore (LF), right hind (RH), right fore (RF). As shown in the lower left schematic (which indicates foot placement before and after the LF footfall), this sequence gives a large triangular polygon of support for most of the step cycle (the position of the center of mass is indicated by a filled circle). Primates, by contrast, use a diagonal sequence gait (lower right), which gives a much smaller polygon of support, but ensures that the main supporting leg (RH here) lies close to the center of mass as the opposite forelimb (LF) is planted. As a result, should the support give way, a primate can easily save itself from a fall.

maintained throughout the step cycle (Fig. 2-4). Primates and the grasping marsupials, on the other hand, use a *diagonal* sequence gait, in which a left/right alternation of the hindlimbs is matched with a delayed *right/left* alternation of the forelimbs: the sequence is right hind, *left* fore, left hind, *right* fore. At first glance this sequence makes no sense, for the polygon of support is reduced to a mere sliver. However, on a thin branch, the large support areas of a lateral sequence gait can't be realized. Worse, when a hand makes contact with a branch at the beginning of its stance phase, the opposite supporting foot is at full stretch when in lateral sequence mode, so if the grasped branch were to give way, the animal would likely find itself way off-balance. In a diagonal sequence gait, the supporting foot is further forward—nice and close to the center of mass—so if a grasped branch turns out to be flimsier than it looks, the animal can easily tilt back and avoid a fall.

This adoption of the diagonal sequence gait for health and safety reasons holds the key to the whole primate story. For simple geometric reasons, switching the footfall order causes the hindlimbs to support a more substantial fraction of body weight throughout the step cycle than in a lateral sequence gait. This shift in weight-bearing responsibilities had a number of important evolutionary consequences. For one, given the right selective pressures (e.g., a further increase in size), it's a short step from putting a bit more weight on the back legs to going orthograde. This is not something that applies only to the apes, by the way: throughout their history, prosimians have shown a consistent tendency toward orthogrady, which has in some cases given rise to some familiar-looking locomotory behaviors, such as the upright bipedal skipping used by grounded Madagascan Verreaux's sifakas.

Of equal significance to an orthograde tendency, saddling the hindlimbs with the lion's share of the weight-bearing task, when combined with the presence of opposable thumbs, enabled primate forelimbs to become uniquely specialized for manipulation and food acquisition. New foraging methods, involving hand-eye coordination rather than snout-mouth, were only the beginning. The vast majority of primates lead complex social lives, where success relies on knowing your place and who your friends are. Manual grooming is the standard social "cement" of primate comradeship, no doubt because the precision with which primate hands can be operated makes them unusually effective tools for removing parasites, such as lice and fleas. Manipulative hands may thus have opened the door to increasingly complex group dynamics in the primates. This is no small matter: it has long been known that primate group size is strongly correlated with the relative mass of the brain, which seems to indicate that the chief driver of primate brain expansion was the need to keep track of who's who, who's friends or enemies with whom, who lies where in the dominance hierarchy, who owes whom a favor, and suchlike. Our other notable cognitive skills, such as language and physical problem solving, may well be a side effect of this social IQ. If that were not enough, the opposable thumb of course was and still is the linchpin of human technology. All of this was made possible by an enterprising tree shrew–like creature, who with little more than a subtle alteration in the timing of its footfalls ventured boldly where no mammal had gone before: onto the flimsiest branches of the Late Cretaceous trees.

----------

Our exploration of our home group is all but done. But as we bid them farewell, the primates have one further puzzle for us. Of the various tree-loving groups of animals, primates are decidedly unusual in their complete lack of gliding representatives. Gliding squirrels have evolved several times independently, as have gliding lizards and frogs—there are even a couple of gliding snakes. There are many marsupial gliders—feathertail, sugar, and greater gliders, to name but three. Strangest of all, our own close relatives the colugos are possibly the most accomplished gliders alive today. Something about the primates seems to have closed the door to the aerial realm, despite the great benefits that a life on the wing might offer. In the next chapter, we'll find out what that something may be, and much more besides, as we look into that most enviable of locomotory transitions: the origin of flight.

3

# Leaps of Faith

## In which we find out why evolving flight is so dreadfully difficult

The boy began to rejoice in his audacious flight
and deserted his leader, and drawn by a desire for the sky
he pushed his path higher. The nearness of the sun quickly
softens the fragrant waxes, the feathers' binding;
the waxes have melted: he shakes his bare arms
but lacking wings he secures no air,
and shouting his father's name his face
is swallowed by the azure water...

—Ovid, *Metamorphoses*

With that, Icarus's brief flight is brought to an abrupt and permanent end in the Aegean Sea, to serve as a warning to all overconfident youngsters that it might be a good idea to listen to your dad once in a while. By this token, the tale is but one of many fables of the perils of overreaching that pepper ancient literature and folklore, from Arachne the weaver to the Tower of Babel. But for such a short story (just a few lines in an ancient Greek compendium before Ovid padded it out in his *Metamorphoses*) Icarus's fate has an unusually strong resonance. To my mind this can only be because it encapsulates our ultimate and, until recently, impossible dream: to throw off our earthly shackles and take to the air. It isn't the only example of such an allusion: the mythologies of most cultures are littered with evidence of our aerial fantasies in the form of multifarious gravity-defying humanoids, from angels to witches.* Indeed, heaven itself, or the nearest culture-specific equivalent, is traditionally regarded as being "up there," giving a divine spin to our wishes.

All this could be regarded as a case of the grass's being greener on the other side, for flight is the only principal locomotory technique we can't do under our own steam. But there's clearly more to it than that. For one, the unrestrained dynamism of self-guided flight offers an undeniable thrill to anyone lucky enough to have the opportunity to experience it. I'll be taking a look at this sort of thing in Chapter 10. For now, however, I'm going to concentrate on the implacable realities of Darwinian fitness. From this point of view, flight is a no-brainer. While it can be rather expensive as far as power consumption is concerned, in terms of the cost of transport (the amount of fuel it takes to travel a given distance), it's a vastly cheaper alternative to making the same journey on foot. Flying is also patently faster than walking, which can assume real importance if competitors are racing toward the same patch of food. And it's safer: earthbound predators usually stand little chance of catching flying prey, who have access to an entire extra dimension of escape routes. Finally, the potential to travel long distances over any terrain opens up the world. Consider arctic terns, which fly from pole to pole twice each year, and so never need to weather the hardships of winter. On a smaller scale, the most fleet-footed runner is brought to a halt by a river, but an aeronaut can keep going as long as there's air.

Given all that, the chief mystery surrounding the evolution of flight isn't why the lucky few can fly—it's why everything else doesn't. Only for aquatic animals is the answer completely straightforward. Quite apart from the fact that many of them never come anywhere near the air, swimming generally has an even lower cost of transport than flying, so water dwellers rarely have anything to gain from aerial experimentation. The exceptions are flying fish and flying squid, which use fin/tentacle-supported glides to effectively teleport away from underwater predators. Every terrestrial creature, however, should be set to receive all the benefits mentioned earlier, yet true powered

* As a Brit, I feel I ought to mention the legend of King Bladud (pronounced "bladdered"), who contracted leprosy, cured himself by bathing in restorative mud after watching his pigs do the same, founded the city of Bath on the site, then used the dark arts to grow a pair of wings and take off across the country, only to crash-land messily in London. His son is better known: Shakespeare's King Lear.

flight has arisen only four times among them: in the insects, birds, bats, and extinct pterosaurs. Admittedly, gliding has evolved many more times than this, and a passive use of air currents is fairly commonplace for the tiny, but that still leaves many animals firmly tied to the ground. For some reason, in spite of all the blessings it brings, flight has proved an astonishingly difficult nut for evolution to crack. To find out why, we'll first need to investigate the forces required to ride the wind.

## BALANCING ACT

Any object in a gravitational field is doomed to accelerate earthward unless it experiences an upwardly directed force equal to its weight, so the key challenge of flight is how to generate this force in midair. This would be no problem at all if we had access to buoyant bags of lighter-than-air gas, but as nature has yet to discover such aerostatic tricks, we're left with aerodynamic forces—those that appear only when moving relative to the surrounding air. Such forces are conventionally resolved into two components: drag, which acts parallel to the direction of motion (thereby opposing said motion), and lift, which acts perpendicular to it. If you're relying on drag alone, you must descend with respect to the air—this is the parachute principle—but if instead you can generate enough lift to balance your weight and enough thrust to balance your drag, purely horizontal movement is all that's required, with no loss of altitude. This is how powered flight works (Fig. 3-1a). If you can produce lift but not thrust, your movement must have both horizontal and descending vertical components—in other words, you must glide at an angle to the horizontal, the steepness of descent being determined by the lift-to-drag ratio (Fig. 3-1b–c). In this case, the vector sum (or resultant) of the lift *and* drag balances your weight.† We'll deal with the simple, drag-only, parachuting situation first.

We can get some idea of the magnitude of drag in various circumstances by turning to Newton's second law of motion, which tells us that a force $F$ applied to an object is equal to its mass $m$ times its acceleration $a$. For a moving fluid (which includes air), mass can be a tricky thing to define, so we're better off replacing this term by density $\rho$ (Greek letter rho) times volume $V$,

† *Vector sum* means the directions of these forces must be added as well as their magnitudes. Adding vectors is rather like working out a shortcut from a series of individual directions: if you're told to go 5 miles northwest, then 5 miles northeast, you'll reach your destination faster if you just go 7 (and a bit) miles north instead. That 7 miles north is the vector sum.

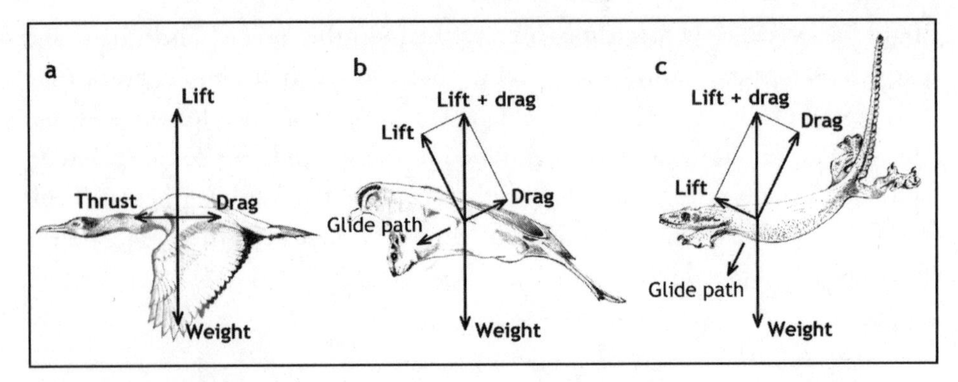


3-1: The balance of forces in powered and gliding flight: for a powered flier (a), lift—always defined perpendicular to the relative air velocity—balances weight, and thrust balances drag—always defined parallel to the relative air velocity. When gliding (b, c), the flight path tilts such that the total aerodynamic force—the sum of lift and drag—balances the weight. When the lift-to-drag ratio is high (b), the glide path is shallow, but if low (c), the path is necessarily steep. If there is no lift, drag alone must balance weight and the animal descends vertically through the air.

which can itself be replaced by length $l$ in the direction of travel times the area perpendicular to this direction $A$. Similarly, we can replace acceleration by the velocity $v$ divided by time $t$. So, the force experienced by a mass of air as it's pushed downward by a falling object, which, by Newton's third law, will be equal and opposite to the drag force experienced by said object is:

$$F = \rho \times A \times v \times \frac{l}{t}$$

As $l/t$ is velocity, this simplifies to:

$$F = \rho \times A \times v^2$$

which was what Newton himself came up with. However, this turns out to be a whopping overestimate, for it assumes that the air ahead of the falling object is unable to get out of the way, instead piling up ahead of it like ice cream in front of a spoon. In reality, air flows around the edges of a plummeting item to fill the void behind. Nevertheless, the basic relationship holds: drag is some fraction of air density times area times the square of the velocity.

That fraction—called the *drag coefficient*—depends on the orientation and shape of the object; in particular, how streamlined it is. So:

$$D = \rho \times A \times v^2 \times C_D$$

$D$ is the drag force, and $C_D$ the drag coefficient. Now, an object in free fall will accelerate toward the ground until the upwardly directed drag force is equal and opposite to its downwardly directed weight force. From that point on, it falls at a constant velocity—the so-called terminal velocity—which can be found by simply substituting the object's weight $W$ for drag in the above equation and rearranging:

$$v = \sqrt{\frac{W}{\rho \times A \times C_D}}$$

Hence, if density and the drag coefficient remain constant, the terminal velocity of an object depends only on the ratio of its weight to its surface area.

Now we're in a position to think again about the harsh realities of life off the ground. Terminal velocity is something that natural selection would be expected to try to reduce for any organism liable to fall from a reasonable height: to avoid injury, this maximum speed should be as low as possible. That means maximizing area and minimizing weight. Small organisms are at an inherent advantage here, because they're naturally high-area, small-weight objects. As size increases, surface area goes up with length squared, whereas weight goes up with length cubed. All things being equal, terminal velocity therefore increases with size, and as the all-important kinetic energy of impact depends on mass times velocity squared, larger organisms are effectively excluded from the incipient fliers club: big creatures need a lot of extra area to save themselves from a fall,* and the chances that a single mutation will provide such a safety net overnight are negligible. Conversely, however, if a creature is too small, its kinetic energy at impact won't be sufficient to

* No one has summed up this situation quite as well as J. B. S. Haldane in his 1928 essay *On Being the Right Size*: "You can drop a mouse down a thousand-yard mine shaft and, on arriving at the bottom, it gets a slight shock and walks away. A rat is killed, a man is broken, a horse splashes."

cause any kind of damage at all, giving little evolutionary incentive for further change.

Or so we might think. Further reductions in terminal velocity are actually far from useless for the tiny. Remember, we're talking about the speed of descent relative to the surrounding air, and air is rarely still. Should it rise faster than an organism falls through it, altitude will be gained. That in itself is nothing special—going straight up will do you little good beyond giving a decent view of the surroundings—but superimpose horizontal air movements and you have all the makings of a wonderfully cheap if unpredictable means of traveling from place to place. If you're very tiny, this natural conveyor belt can be exploited with no modifications at all. Fungal spores fall into this category. Get a bit larger and additional area proves increasingly useful, although as long as you stay reasonably small you can get away with surprisingly little. Many kinds of infant spiders ride the wind to disperse, using the extra drag from a few silken threads (the fabled gossamer) to take them aloft.

At vertebrate size, though, a bit of silk isn't going to do anything to reduce terminal velocity. What would be useful at this scale is something approaching an actual parachute: a thin membrane that adds lots of area but little weight. When stretched taut between supporting bones, skin fits the bill perfectly, so it's no surprise that two of the three vertebrate powered fliers and nearly all the thirty or so vertebrate gliding lineages have independently turned to their skin as a means of balancing their weight. You don't need much to start you off as long as you're fairly small: the "flying" frogs of the American and Asian tropics, for instance, make do with a bit of webbing between the fingers and toes.

These modest aerodynamic accoutrements seem pretty basic, but there's more to such webbing than meets the eye. Were flying frogs to keep their mini-parachutes perfectly level in free fall, the only thing they'd get out of them is drag, which as we've seen is no bad thing, although in the stifling air under the rain forest canopies of their homelands, they're unlikely to extend their horizontal range by merely slowing their descent. At the other extreme, if they decided to go for a headfirst nosedive, they'd get nothing at all from their webbing. Between these two orientations—parallel to the airflow and perpendicular to the airflow—wondrous things may happen. As long as the angle between the airflow and the webbing isn't too large, a force appears that acts perpendicular to drag—this is lift, of course, and once it's on the scene, we're into gliding territory.

## THE ULTIMATE PICK-ME-UP

Lift must be one of the most misunderstood phenomena in mesoscale physics (the scale at which we can ignore relativity and quantum mechanics). Despite the considerable use we make of it these days, there's little agreement, even among the experts, about where lift comes from, in physical terms. The mathematical basis is understood extremely well, all things considered, but provides only a *description* of flow and its associated forces, not a true explanation, because at no point do the relevant equations distinguish cause from effect. If you're not careful, it's easy to get the two the wrong way around, as exemplified by the abuses to which Bernoulli's principle has been subjected in attempts to explain lift. I'll be coming back to such abuses shortly. First, however, we need to get to grips with the physics. In essence, Bernoulli's principle is a fluid-specific expression of the law of the conservation of energy, stating that the sum of the kinetic and potential energies of a moving fluid must remain constant. Kinetic energy depends on a fluid's velocity, whereas potential energy depends on its pressure,* and a change in either will be accompanied by an equal and opposite change in its counterpart. That makes complete intuitive sense—it's easy to appreciate that fluid flows from high to low pressure: in other words, that a drop in pressure is accompanied by a rise in velocity.

That much is fine. However, the most common explanation of lift reverses the comfortable logic. It usually goes something like: air flows faster over a wing than underneath it, *causing* the pressure to drop topside (because that's what Bernoulli says), and the resulting pressure difference lifts the wing. The principle is supposedly demonstrated by a simple armchair experiment that you may perform at your leisure: blow over a strip of paper and it rises. It's pretty convincing, if you ignore the troubling question of *why* air flows faster over a wing when there are no lungs at hand.† But even if you do, it doesn't take long to find the plot holes in the story. Try the paper-levitating experiment again, but this time, instead of holding the strip only at the end nearest your mouth, support it with your open fingers to keep it flat. This time there's

* Specifically, its *static* pressure. *Dynamic* pressure—what you feel when air blows in your face—is another way of describing a fluid's kinetic energy.

† A particularly heinous pseudo-explanation for this phenomenon, which can still be found in the odd textbook, is that air molecules separated by the leading edge of a wing must reunite at its trailing edge. Because the high road is longer, so the story goes, the air must therefore travel faster over the top. This is nonsense: with a wind tunnel and a means of periodically and regularly introducing colored smoke into the airstream, it can be quickly and easily demonstrated that the upper airstream reaches the trailing edge well ahead of the lower.

no lift, and were you to measure the pressure of the blown airstream, you'd find no significant deviation from ambient.

So, what's going on? Logic suggests that the generation of lift must have something to do with the curvature of the paper, as this was the only thing that was different in the two versions of the test. This is indeed the case, although you'd need to look very closely to see why. Imagine you were to stand on that strip of paper as air was blown across it and then shrink down to such an extent that you could discern individual air molecules. From your vantage point, perched on the craggy molecular landscape of the paper, the flow overheard resembles a pitched aerial battle. While the overall wind direction is readily apparent, the air molecules are constantly colliding, knocking one another toward or away from the paper. Where the surface is flat (well, flat at the macro scale), these molecular movements to and from the surface are balanced, but if you were to hike to where the paper curves away, you'd see that most of the molecular trajectories were directed away from the surface. At our usual scale, that local depletion manifests itself as a pressure drop: the source of lift.*

In my opinion, this Bernoulli-free explanation of lift is considerably easier to grasp than the standard version you'll find in textbooks, and it does a much better job of accounting for familiar aerodynamic phenomena. We can now explain why air flows faster over the upper surface of a wing: the low pressure there simply causes the local flow to speed up (invoking Bernoulli's principle, but the right way around). Increasing the overall flight speed increases the rate at which near-surface air is depleted, giving a lower pressure and greater lift: specifically, lift goes up with the square of the relative air velocity, just like drag. Increasing the curvature (camber) of the wing or the angle between it and the oncoming airflow (the angle of attack) has a similar effect up to a point, because the molecular trajectories tend to be more sharply divergent from the wing surface. These shape/orientation effects are quantified as the *lift coefficient*—analogous to the drag coefficient. The upper limit of the lift coefficient exists because if the molecular trajectories diverge too sharply there's nothing to stop the near-surface air downstream reversing direction to fill the void. This is what's known as a stall, marked by increased drag and diminished lift.

* The low pressure over the wing applies a centripetal acceleration to the airflow, deflecting it downward. Lift can also be explained as the reaction to this downward deflection of air, invoking Newton's third law.

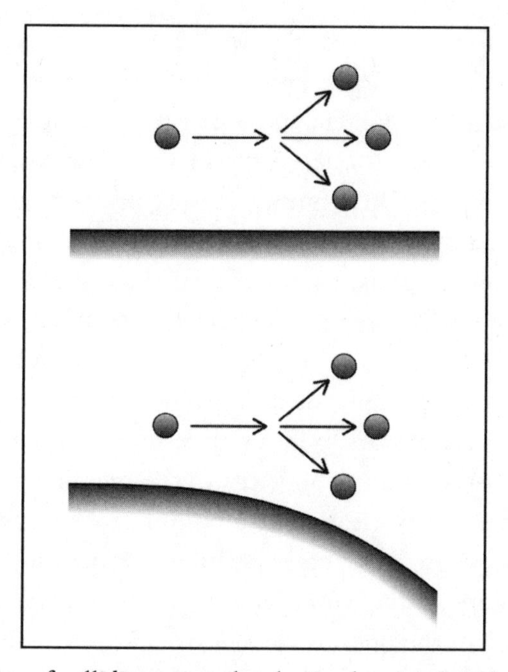

3-2: Trajectories of colliding air molecules in the vicinity of a solid surface. When the surface is flat and aligned with the relative wind direction (top), as many air molecules are knocked toward the boundary as are knocked away from it, but should the surface curve away (bottom) there will be a net export of air molecules from the immediate vicinity, causing a pressure drop.

## GLIDE PATHS

What does all this mean for an animal's aerodynamic career? Well, the lift provided by a modest set of wing membranes, as modeled by flying frogs, can give a smallish creature an assortment of juicy selective benefits as long as said creature is well stocked with gravitational potential energy. In practical terms, this means it needs to be up a tree. By adding some horizontal range to a descent even in the absence of a side wind, a bit of lift might allow an animal to move quickly about a forest without incurring the expense or risk of clambering down to ground level, hiking across the predator-packed forest floor, and scrambling all the way back up again upon reaching the desired tree. Furthermore, there's continual scope for gradual evolutionary improvement: if the glide angle can be made just a bit shallower, less height will be lost for a given horizontal excursion, and less energy and time will be spent clawing it back; conversely, the range for a given drop will be extended.

To flatten the glide angle, an animal needs to increase its lift-to-drag ratio (see Fig. 3-1), which can be achieved in the first instance simply by enlarging the protowings. Exactly how this is done depends on where the wings are located on the body and what they're made of. Some gliding frogs have unusually long fingers and toes with expanded webbing, whereas gliding lizards of the genus *Draco* support their membranes with elongate ribs, a "design" also seen in a number of extinct gliding reptiles. Gliding mammals use broad flaps of skin stretched between all four extended limbs. They include flying squirrels, the marsupial sugar gliders, and the colugos that we briefly met in Chapter 2—the gliding membrane of this latter group encompasses not just the arms and legs but also the tail and the long fingers and toes. Note the group-specific signatures of these various wing types. Using ribs as supports appears to be a reptile-only solution, just as stretching a flange of skin between arms and legs is the mammalian calling card. A would-be glider's ancestry clearly has a big influence on where a protowing starts to grow. In no group is this more apparent than in the gliding snakes of the genus *Chrysopelea* (the stuff of nightmares!) whose stripped-down morphology writes off all the options we've looked at so far: the only thing they can do is splay their ribs to flatten their body. The influence of ancestral form on future evolutionary possibilities will prove absolutely critical when we come to explore the origin of powered flight.

An increased lift-to-drag ratio is but one effect of wing area expansion. A more obvious consequence is that for a given speed and lift coefficient, a bigger lift *force* will be produced, with the upshot that the creature's weight can be supported at a lower airspeed (this is precisely analogous to the relationship

**3-3:** Reptilian and mammalian aerodynamic adaptations: the gliding lizard *Draco volans* (left) and a colugo or flying lemur (right).

between drag and terminal velocity for parachuters). Keeping speed down makes landing safer, but that's only the half of it. Immediately following takeoff, there's a brief phase of acceleration while the aerodynamic forces build to the point at which weight is balanced, when true gliding begins. If a glider wants to keep its height loss to a minimum, its steady gliding speed needs to be as low as possible: the lower this figure, the sooner it will be reached. All the more reason for natural selection to grow those wings.

So far, it appears that becoming a glider is simplicity itself: as long as you're up a tree and small enough for stubby protowings to provide significant aerodynamic forces relative to weight, the evolutionary path seems clear. But this can't be true, or else every modest-size tree dweller on the planet would be a glider, including plenty of primates, as I noted in Chapter 2. The thing is (and the woolly opossum experiment described in Chapter 2 makes this very clear), you don't have to be a glider to avoid making dangerous and costly trips across the ground when going from tree to tree. If you're reasonably adept at navigating in the canopy—and primates are exceptionally good at navigating in the canopy—chances are you'll be better off walking or jumping to the nearby branches of a neighboring tree than taking a leap of faith to its more distant trunk. Until your protowings have grown fairly large, such leaps will be so precipitous that, even if you're not a primate, you'll waste far more energy climbing back up afterward than you will have saved by eschewing the high road in the canopy. Natural selection is unlikely to look favorably upon such an unnecessarily grueling lifestyle.

Surprisingly, then, the prospect of improved locomotory economy is all but ruled out as an incentive to start gliding, simply because the prospect is too distant. What other selective factors might be pertinent? One promising possibility goes back to the idea of predator evasion: not terrestrial this time, but the enemy within—the tree-dwelling predators with whom the incipient gliders live cheek by jowl. Jumping out of a canopy is an effective way of getting away from the jaws of death, but it would be nice if the escape wasn't immediately followed by a long drop to ground level where, even if the impact or terrestrial predators didn't finish you off, you may be faced with a long walk home. A bit of lift can come in very handy in such circumstances, as indeed it will if the fall was merely caused by a plain, old-fashioned slip. Getting a decent horizontal range really doesn't matter in this context—all the protoglider needs to do is generate aerodynamic forces sufficient to nudge its descent trajectory, so that it can swoop back to the tree it fell from. Early indications are that this benefit may well have kick-started the many acquisitions of gliding behavior. For instance, when Sharon Emerson of the University of Utah and Mimi Koehl from the University of California, Berkeley, tested flying frog models in a wind tunnel, they made the surprising discovery that the real animals weren't getting the best possible lift-to-drag ratio from their aerodynamic gear. Models with a stretched-out posture gave a shallower glide angle than did those in a lifelike tucked-in configuration, but had poorer control characteristics. Controlling their descent must be more important for these creatures than overall horizontal range.

Further evidence for the evolutionary primacy of aerodynamic control comes from the observation that many small tree dwellers use simple postural adjustments to alter the trajectory of a fall despite lacking even the stubbiest of wings. For example, tropical biologist Stephen Yanoviak recently noticed while working in the rain forests of Panama that canopy-dwelling ants, which suffer falls at an alarming frequency, rarely reach the ground following such a mishap. They instead use the aerodynamic forces generated by their wingless body to direct their descent back to their home tree. Several lizards and frogs, despite their larger size, are known to do the same sort of thing. Such skills could be gradually honed by the addition and subsequent expansion of protowings, with every enhancement giving selective returns, thanks to the energy saved, until the stage is reached when a trainee glider is finally able to reap the economic benefits of effective aerial tree-to-tree travel. Which, of course, brings us right back to our earlier question: why did natural selection take the no-winged to protowinged step, and thereby open the door to gliding proper, in only a few cases?

A possible clue to this stubborn mystery is the unusual global distribution of gliders. The vast majority are found in the jungles of Southeast Asia, with another bunch—mainly marsupials—in the eucalyptus forests of eastern Australia. These forests are notable for their height, but also for the cathedral-like openness of the spaces beneath their high canopies. This is important because even expert gliders lose a fair bit of height at the very beginning of each swoop, as they accelerate to their minimum steady gliding speed. That means that if gliding from trunk to trunk is to beat clambering from canopy to canopy in economic terms, the horizontal range of the glide must be relatively long. The cavernous forests of Southeast Asia and Australia are just the ticket: the relatively high probability of long drops favors the evolution and refinement of midair control, and the clear space, when coupled with the height, gives gliders what they need to reap the full benefit of their aerial prowess. When all's said and done, it thus appears that the critical determinant of whether the evolutionary path to gliding is available is simply being in the right place at the right time. Unless, that is, you happen to be a primate. Even with the extra incentive provided by the Southeast Asian rain forests, primates have remained resolutely implacable in their distaste for aerodynamic experimentation. Only gibbons come vaguely close with their exhilarating brachiation (and it's probably no coincidence that they are found in Southeast Asia). Perhaps we primates are just too good on fine branches to ever make incipient gliding worthwhile, not to mention the trouble a primate protoglider would have in trying to land without claws on a tree trunk, or on a narrow branch while having to keep its arms outstretched. Those opposable thumbs turn out to be something of a mixed blessing.

## GETTING IN A FLAP

The main aim of gliders is to lose as little height as possible. The dream is to lose none at all, of course, but there are certain tedious constraints that make this goal difficult to achieve. Crucially, drag cannot be reduced to zero even with the best streamlining in the world, so the total aerodynamic force always tilts backward relative to the path of the wings. Gliders must rotate their trajectory to compensate, and so cannot help but head groundward. There is, however, a loophole. Nowhere is it written that the aerodynamic force must necessarily tilt back with respect to the *body's* path through the air. If the airstream strikes the wings from beneath this path, the total force may end up leaning forward, giving both lift to balance weight and thrust to counteract the overall drag. On the face of it, this is a wonderfully simple trick to pull off: the wings just have to be moved downward in flight. The forward velocity of the animal and the downstroke velocity of its wings will sum to give the oblique airflow it needs. This, of course, is the basis of powered, flapping flight (Fig. 3-4).

You can probably guess what's coming next. If it's so easy to do, why has it only evolved four times, despite the great multitude of gliders? Well, for a start, any benefits of flattening the flight path must outweigh the energetic costs of flapping the wings, although given the expanded foraging radius opened up by flapping, this doesn't sound like a difficult barrier to negotiate. What's more important is that effective flapping isn't as easy as I just made out. Downstrokes can't last indefinitely, of course; at some point the wings must reverse their direction and reset. But if an upstroke is merely a mirror image of the downstroke, all the downstroke's good work will be undone:

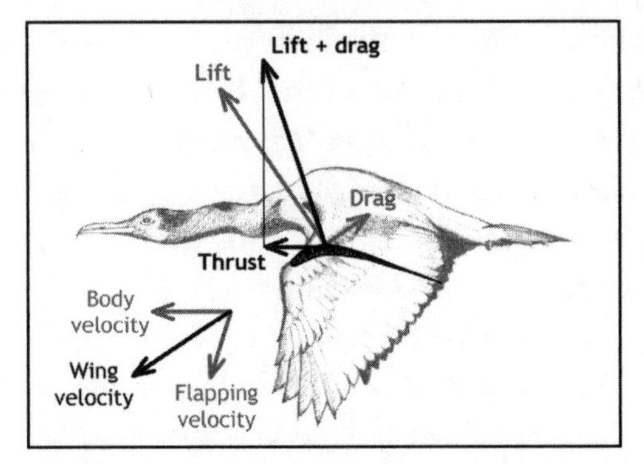


**3-4:** How flapping generates thrust: during the downstroke, the total velocity of the wing (body velocity plus flapping velocity) is inclined with respect to the flight path, so the lift and drag vectors will also be inclined by precisely the same angle. As long as the lift-to-drag ratio is sufficiently high, the total aerodynamic force tilts forward, providing thrust.

because the airflow now strikes the wings from *above* the body's trajectory, the total aerodynamic force tilts further back than usual, canceling out the thrust of the downstroke. To flap effectively, the wings must be modified during the reset so as to reduce the total force developed during the upstroke, usually by reducing the angle of attack. This is harder than it sounds. Big changes to the angle of attack of the wings alone—independent of the orientation of the body—require that they have a fair bit of rotational flexibility, which isn't easy when the base of each wing is broad, as in most gliders. Then there's the fact that unless the lift-to-drag ratio is high to begin with, an animal will have to flap extremely hard to crank the total aerodynamic force sufficiently far forward during the downstroke to get thrust, making the transition more trouble than it's worth.

This is the big problem, for truly high lift-to-drag ratios are hard to come by. A simple expansion of the wings suffices early on, but once they've reached a decent size, the trend can only be continued by making them long and thin (in Chapter 4, you'll learn why this works). Such changes are also necessary to gain independent control over the angle of attack of the wings. Whether the transition from gliding to flapping is possible thus crucially depends on whether the anatomy of the wings is conducive to the required elongation, quite apart from the fact that the supporting structures and their muscles must be able to withstand and deliver the vigorous pumping that powered flight demands. Few of today's gliders meet these criteria. It's difficult, for instance, to see how the rib-supported wings of the gliding lizard *Draco* will ever acquire the range of movement and muscle power needed to take the next step—the design is an evolutionary dead end. But that's the trouble with natural selection: it deals only with the concerns of the present and cannot look ahead. The four groups of powered fliers were the lucky few whose ancestors just happened to have wings that ultimately proved flappable.

So, what were their respective secrets? The clues are there in their present-day anatomy, if you know how to read the evidence. Bats, for instance, support their wings with their elongated fingers (the thumbs remain free), so their gliding ancestors must have had webbed hands on top of the standard mammalian-type wing membranes. I would hazard a guess that those proto-bats were much like today's colugos—the nearly-but-not-quite-lemurs whose gliding membranes encompass their entire body. With their fingers embedded in the wings, the gradual elongation that was required to improve the lift-to-drag ratio could be achieved without any risk that the supporting structures would become too flimsy or the membranes uncontrollable. Quite the opposite: the moment-by-moment control that bats have over their wings is exquisite. As a result, their maneuverability, especially that of the echolocating (sonar-guided), insectivorous "microbats," is exceptional—it has to be, given the aerial prowess of their preferred prey.

Pterosaurs—who at 220 million years of age were the first vertebrates to take to the skies—provide an illuminating contrast, for with their membranous wings they can seem batlike at first glance. However, while bats support each wing using four of their five fingers, pterosaurs did it all with one—a grotesquely elongated fourth finger. We can therefore surmise that their gliding ancestors lacked colugo-like hand webbing. Instead, the wing membrane must have attached only to the last, fourth finger of each hand (presumably the pinkies had already been lost). When the time came for the wings to elongate, it was this digit alone that was stretched by natural selection, with important consequences for the later aerodynamic career of the group. Lacking the additional membrane-spanning fingers of bats, the pterosaur wing was more difficult to control, but was arguably better suited for the sort of extreme lengthening and narrowing seen in today's albatrosses and frigate birds. What pterosaurs lacked in agility, they made up for in aerodynamic efficiency, and in the sheer size of their wings, which in the largest species spanned up to 40 feet.

Insects—our third group of flappers—started flying long before any of their vertebrate protégées. The great 300-million-year-old Carboniferous coal forests were home to many different kinds—including giant dragonflies with wingspans measuring nearly 20 inches—and 410-million-year-old Devonian rocks from Scotland have yielded fossilized mouthparts that look like those carried by today's winged insects. Insects' take on flight is somewhat different from what we've seen in the bats and pterosaurs, partly because most are much smaller, but also because, as arthropods, their wings are derived from folds of the exoskeleton with no direct connection to any preexisting limbs. From the first, the wings therefore had an inherent structural integrity—they didn't need to be stretched out like skin wings—so natural selection was able to elongate them and render them independently adjustable with little fuss.

Do not think, however, that flapping flight was a straightforward evolutionary prospect for the protoinsects. True enough, substantial aerodynamic forces (relative to weight) are easy to produce when you're small (remember the plummeting Panamanian ants); but that's part of the problem: if your descent can be controlled perfectly well without saddling yourself with extra sets of costly, cumbersome appendages, then why bother? As a further disincentive, high lift-to-drag ratios are particularly elusive for small fliers (for reasons I'll divulge in Chapter 9), making a transition to longer-distance gliding difficult. Both issues would have been mitigated if the first insect test pilots were at the larger end of the current size range—model tests indicate that at a body length of 4 inches or so, stubby winglets would have significantly reduced the glide angle, with further reductions accruing steadily as the winglets elongated. The fossil evidence of early insects is patchy in the extreme, but it doesn't rule out such sizes. A more serious obstacle is that winged insects seem to have predated trees. The tallest plants of the early Devonian—when those fossilized mouthparts crop up—were spindly shrubs no more than 3 feet tall. It's hard to imagine flight evolving from such stunted launch pads. However, plants didn't have a monopoly on height back then. Believe it or not, the tallest known organisms at the time were fungi called *Prototaxites*, which grew to the dizzying heights of 26 feet or more. Before your imagination runs away with itself, they were simple, pillarlike growths, not giant mushrooms; nevertheless, they must have been an impressive sight, towering over the contemporary plants like the ancient megaliths of a long-dead civilization. Who knows? Perhaps they and their like gave the protoinsects the leg up they needed to get their aerial experiments under way.

Aside from their exoskeletal composition, insect wings differ from their bat/pterosaur counterparts in one other vital respect. Whereas the vertebrates have or had two wings apiece, almost all flying insects have four (although one pair is often modified beyond recognition). Thus there is a measure of redundancy to the insect flight apparatus: under appropriate selective pressures, one pair of wings can be tweaked without screwing up the whole system. There's no shortage of examples of such tinkering. In beetles, for instance, the front pair of wings have been converted into a hard, protective shield under which the conservative back pair are stowed when not in use. In flies, the back pair have shrunk into drumsticklike gyroscopic devices (halteres) that provide detailed in-flight motion data. Flies' exceptional swat-evading agility is largely attributable to these structures. The evolutionary freedom provided by the duplicate wing set has undoubtedly been a major contributor to the staggering diversity of insects. Remember this trick—we'll be seeing it again.

As far as we can tell, and despite their radically different starting points, the discovery of flapping flight by the ancestors of bats, pterosaurs, and insects probably played out in much the same way each time, with a controlled descent phase and a progressively refined gliding stage preceding the thrust-generating breakthrough, all ultimately fueled by gravitational potential energy. But what about our final group of flapping fliers? Did the birds follow the same well-flapped path?

## FEATHERED FRIENDS

The idea that birds are modified carnivorous dinosaurs is fairly well known now, thanks in no small part to the success of the film *Jurassic Park*. The proposition, however, is actually a very old one: Thomas Henry Huxley, Darwin's staunch champion, drew attention to the resemblance between dinosaurs and birds back in 1870. The idea fell out of favor in the early twentieth century, but was revived by Yale University's John Ostrom in the 1960s and '70s. He compared the famous *Archaeopteryx* from the Late Jurassic with the Cretaceous dinosaur *Deinonychus* (which was in *Jurassic Park*, but under the name of the considerably smaller *Velociraptor*) and found many shared characteristics, particularly in the wrist and hand. The proposal has since been nearly comprehensively settled by the discovery of a large number of dinosaur fossils in northeastern China that, like *Archaeopteryx*, still carry the preserved remains of their feathers. The consensus today is that birds constitute the sister group (i.e., closest cousins) of the deinonychosaurs, a group that includes *Deinonychus*, *Velociraptor*, and their relatives. The two groups are collectively known as the paravians. Slightly more distant relatives of the birds include the ostrichlike ornithomimosaurs and the perennial favorite *Tyrannosaurus rex*. On the basis of the available evidence, all of this lot were probably feathered to some extent, which may come as a shock to *Jurassic Park* fans.

It goes without saying that the standard pathway to powered flight, with an intermediate gliding phase, requires that preflappers have access to plenty of gravitational potential energy; this means they have to be tree dwellers. However, looking at the birds' closest relatives, we seem to be faced with a host of relentlessly terrestrial, fast-running predators, with not a tree hugger in sight. John Ostrom was well aware of this conflict, and could see only one way out: he proposed that powered flight in birds evolved from the ground up, with no tree-dwelling ancestors and no parachuting or gliding preliminaries—just a seamless transition from running to flapping.

Now, a ground-up origin of flight in bats and pterosaurs was out of the question: because the legs are (or were) needed in flight to keep the wing membranes under sufficient tension, a running takeoff is or was more or less impossible. The only way bats can take off from the ground is by jumping from a standing start. But birds don't have this problem because, like insects, their wings are free of their legs: the relatively rigid feathers don't need to be stretched out to make them aerodynamically competent. Hence, there's nothing to stop them operating their legs in any desired manner regardless of what the wings are doing, so most of them can run and flap at the same time, and most have no trouble taking off from the ground. Of course, that's all well and good once the wings have actually evolved. If starting from the trees as did bats, selection for improved aerodynamic performance can get the protowings going and then gradually expand them in the context of gliding. On the ground that just won't work, so if our bird-in-waiting is to have any hope of fulfilling its dream, some selective benefit unconnected with flight is going to have to get those wings up to a half-reasonable size.

The idea of wings evolving in a nonflying context may seem ludicrous, but in a broad sense, wings are quite simple structures—little more than flat sheets. There are many reasons why such accessories might start to develop, especially if interference with the existing mode of locomotion is minimal, as would have been the case for bipedal protobirds. In fact, such functional shifts are surprisingly common, and we'll be seeing many further examples throughout the book. They can give the impression that evolution has foresight, seemingly preparing a lineage for an activity that will only be realized in its distant future. For this reason, the process is often referred to as preadaptation. Of course, evolution can't really predict the future—preadaptation, like run-of-the-mill adaptation, is simply a result of the generation-by-generation accumulation of favorable mutations. It just so happens that every now and again, this process unlocks new functional possibilities that later become the dominant concerns of natural selection.

If flight was a late addition to the functional repertoire of bird wings, what might the wings' first role have been? Ostrom himself originally proposed that feathered wings were first used as nets to catch flying insects, although he later retracted the idea, realizing that it was a bit far-fetched. It helps if the function you're proposing is seen in at least some living members of the group of interest, which isn't the case for insect-netting. Some other predation-related functions *are* seen, such as mantling (using the wings to hide a kill from competitors) or balance assistance when subduing prey, but neither activity is possible until the wings are quite large.

A more popular theory for the beginning of wings is that they were originally used as display banners to attract mates or assert dominance, with increased mating success providing the necessary wing-expanding selective pressure. And why not? Many birds use their wings for such purposes. The birds of paradise must hold the record for the sheer exuberance of their feathery ornamentation, but more modest displays are perfectly commonplace. What we need is direct evidence that protobirds used their wings in this way. Conspicuous coloration would likely indicate some kind of display function (not necessarily to the exclusion of other possibilities), but surely we've no chance of finding pigments in 100-million-year-old feathers? Or have we? As it happens, electron microscopic analyses of the ancient plumage of some feathered dinosaur fossils have found tiny (as in, nanoscale) near-spherical particles that have been interpreted as melanosomes—intracellular melanin-filled compartments that are key components of the bird coloring kit. The melanosomes of the Early Cretaceous deinonychosaur *Microraptor* (which we'll meet again shortly) are even arrayed in regular stacks, which would have given its feathers an iridescent sheen.

So far, so good. But while wings alone might be all an arboreal animal needs to set itself on the pathway to powered flight, any terrestrial creature hoping to follow suit would have needed to do much more. To get any aerodynamic benefit from its wings, a ground-dwelling dinosaur would have to have flapped them, and I'm not talking about the sort of leisurely strokes that, say, seagulls get away with: they can fly like this because their steady flight speed is relatively fast—much faster than the sort of speeds a running protobird could attain.* To take off from the ground without such rapid forward motion to do the lion's share of lift production, the wings have to be flapped hard, and I mean *hard*. There's no room for half-measures here: if you can't reach the required flapping speed, your wings will be aerodynamically useless—worse than useless, in fact, because those outstretched feathered arms will add a fair amount of drag, not to mention the cost of the futile flapping itself. Why would a protobird bother? This is a big problem; natural selection, brilliant craftsman though it is, cannot bring about a transition if at any stage along the way the fitness declines, no matter how beneficial the end point. If this were not the case, every terrestrial creature would be able to fly. It therefore seems that the only way our prospective bird on the ground might have realized its dream would have been to develop a fully refined flight stroke overnight.

This is all beginning to sound rather improbable. But the game's not over yet. Back in 2003, Kenneth Dial of the University of Montana discovered that young chukar partridges often flap their poorly developed wings, not to take off (their wings aren't big enough at this stage), but to generate negative lift, like racing car spoilers, to increase their traction when running up steep slopes. He suggested that this was the context in which winged dinosaurs gradually learned how to fly. However, while it's impressive that these little birds can get useful aerodynamic forces from their stubby wings, they must flap them extremely hard to do so: their wingbeat frequency is around ten beats per second, and the wings sweep through an angle of 130 degrees on each beat. Although the requirements are not quite as stringent as when taking off, the idea of a previously flightless dinosaur stumbling upon this behavior still sounds like sitting at a piano for the first time and delivering a virtuoso performance of a Mozart concerto without ever having had a single lesson. It's too big a leap.

We appear to have reached an impasse. As we saw earlier in the chapter, for an animal evolving flight from the trees down, every gradual step toward powered flight can give a selective advantage under the right circumstances. Trying to evolve flight from the ground up, however, is a different matter entirely, requiring such outrageously lucky changes of behavior in the space of a single generation that it seems to be impossible. As Sherlock Holmes

* For instance, *Archaeopteryx*'s maximum running speed, as estimated by bird flight expert Jeremy Rayner, was only about 6.5 feet per second. Gulls generally fly at around 32 feet per second.

would argue, we must hence turn to whatever remains, however improbable. In this case, what remains is the notion that dinosaurs did actually pass through a tree-dwelling phase on the way to becoming birds. But if this were so, where are all the fossil representatives of such a phase?

## SPREAD YOUR LEGS AND FLY

*Microraptor*, as its name suggests, is one of the smallest dinosaurs known, with a tip-to-tail length of about 33.5 inches (most of that being the tail). The first specimen—a beautiful articulated partial skeleton found in northeastern China—was christened in 2000 by the Chinese paleontologist Xu Xing. He recognized it as a diminutive member of the deinonychosaur group (the one with *Velociraptor* in it), but unlike its scarier relatives, *all* the claws of its feet were unusually strongly curved—not just the famous slashing claw. *Microraptor*'s feet were thus more like those of a tree climber than a fast-running predator, although as Xu rightly pointed out at the end of his description, "more evidence [was] needed to confirm this hypothesis."

That was how things stood for a couple of years. Then, in 2003, Xu published a description of a new *Microraptor* specimen. Unlike the first, this one was very nearly complete, but that wasn't its most striking feature by a long shot. Like *Archaeopteryx*, the fossil still sported its flight feathers, but not just on its arms. Feathers of similar size and overall appearance ran down its legs and much of the length of its feet, too. This creature had four wings.

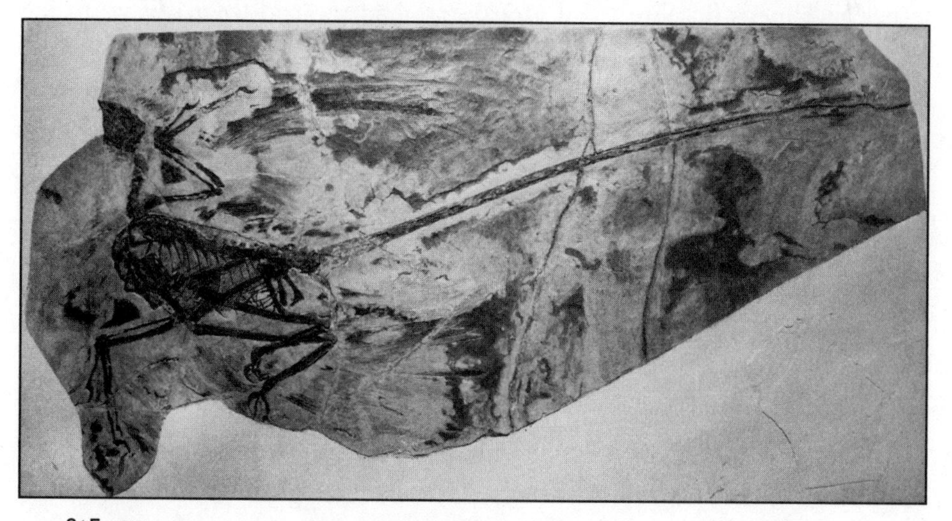

3-5: *Microraptor gui*, a dinosaur with flight feathers on its arms and legs.

The discovery that *Microraptor* had leg-wings was a headline sensation in the paleontological community, although funnily enough a four-winged bird ancestor had been predicted back in 1915 by the visionary American ecologist William Beebe. He came up with the idea after noticing how feather quills developed along the legs of young doves, which he regarded as a last vestige of a formerly four-winged stage. As evolutionary ornithologist Richard Prum put it, *Microraptor* "could have glided straight out of the pages of Beebe's notebooks." Notice he said glided, not flapped. While there's no direct evidence that *Microraptor* couldn't flap its arms, Xu argued that the long leg feathers were incompatible with a terrestrial running lifestyle, reinforcing his earlier claim that *Microraptor* was a tree dweller. It therefore didn't need to flap to get sufficient lift to support its weight. Its existence therefore puts an arboreal, gliding origin of bird flight squarely in the picture, and given the enormous difficulties associated with a ground-up origin, it finally seemed that the birds' pathway to powered flight was similar in most respects to everybody else's.

Since 2003, the evidence in favor of a trees-down origin of bird flight has grown ever stronger, with the discovery of many more dinosaur species with leg-wings, such as *Anchiornis*, *Xiaotingia*, and *Pedopenna*. Significantly, these species are dotted all over the paravian tree, and some, like the three just mentioned, are as old as or older than *Archaeopteryx*. In fact, it turns out that even dear old *Archaeopteryx* had flight feathers on its legs, although they're smaller than *Microraptor*'s: a new specimen described in 2014—only the eleventh to be found—preserves them as clear as day. They'd been overlooked on other specimens because fossil preparators had scraped them away in their eagerness to expose the leg bones. All of this strongly suggests that leg-wings were present in the last common ancestor of all the paravians, which raises the provocative possibility that the likes of *Velociraptor* and *Deinonychus* were secondarily flightless.

All that said, long-cherished theories die hard, and the ground-up camp wasn't going to give up without a fight. And a good thing, too—your opponents will often be more thorough in digging up potential flaws in your argument than your supporters. In this case, the ground-uppers rightly pointed out that we shouldn't leap to the conclusion that the feathered legs of *Microraptor* and friends had an aerodynamic function. Indeed, a standard winglike horizontal posture is highly unlikely, as it would require a most un-dinosaur-like splaying of the legs. However, wings don't have to be fully horizontal to generate lift: even a moderate spread-eagling of the thighs does the trick, and analyses of *Microraptor*'s hip joints indicate a wider range of thigh abduction than is typical of dinosaurs. Furthermore, flight tests of *Microraptor* models carried out by David Alexander and his colleagues at the University of Kansas have shown that those leg-wings fulfilled a vital aerodynamic need. Lacking the compact body of a bird, *Microraptor*'s center of mass lay far behind the center of lift of the relatively narrow arm-wings. Without a posterior source of lift to bring the overall center of lift backward, the creature would have looped the loop and stalled a few moments after takeoff. This isn't usually a problem for protofliers, because skin wings, unlike feathered wings, tend to start off broad.

With so much evidence arrayed against it, it looks like the ground-up theory must finally be consigned to the falsified hypothesis list. Birds learned to fly much as did bats and pterosaurs—in the trees. But that's not to deny their uniqueness. While their overall pathway to flapping flight was like that followed by their aerial cousins, their approach to wing-building had never evolved before and hasn't evolved since. While odd in the extreme, feathers opened up locomotory doors for birds that even the other powered fliers couldn't budge.*

Feathers appear to be solid plates, but in reality they're anything but. Made of keratin (the material that builds our fingernails and hair, as well as reptile scales), a standard flight feather comprises a central or off-center shaft from which branch a multitude of filaments called barbs, each equipped with a complex array of microscopic hooks so that neighboring barbs zip together. The hooks turn what's essentially a collection of branching hairs into a coherent, ultralightweight flight surface that's packed tightly enough to stop air passing through, and strong enough to maintain shape in the face of large aerodynamic forces. We're thus immediately faced with a problem: if the tiny hooks were taken away, the wing would become almost entirely useless in flight, as there'd be nothing holding the barbs together. But if feathers' function is so sensitive to such structural details, how could they have evolved in the first place? In other words, what use is half a feather?

Two hypotheses have been proposed to explain this conundrum. The first, deriving largely from the work of Richard Prum, is a preadaptive scenario that sees in aerodynamically ineffectual fluffy down† the likely primitive

* It could easily have gone differently. A Jurassic paravian, recently discovered in China, has preserved filamentous feathers and, bizarrely, evidence of a skin membrane associated with a unique rodlike bone that articulated with its wrist. Called *Yi qi*, it shows how close the feathered dinosaurs came to evolving in a more batlike direction.

† Actually, it's not completely ineffectual. Pity the barnacle geese of Greenland, which when mere hatchlings have to hurl themselves off precipitous cliffs onto the jagged rocks below

form of the feather. The initially free barbs, so the story goes, later gained the ability to zip together, perhaps in the context of display. In support, several dinosaur fossils appear to carry furlike protofeathers, including the smallish tyrannosaur *Dilong* (what might *that* say about *T. rex*?), with insulation being their most obvious likely function.* Some scientists, however, such as Theagarten Lingham-Soliar and Alan Feduccia, disagree with this interpretation, and think that the preserved filaments are actually the remains of the internal scaffolding of a frill that extended along the back. They favor a scenario in which feathers evolved from elongated reptilian scales that were gradually whittled down into the ultrafine sieves that grace bird wings today. This hypothesis is somewhat consistent with what's understood about how feathers grow: each starts as a curled-up sheet of keratin-secreting cells that later becomes separated into tissue strands, corresponding to the future barbs, by a cellular unsticking process. If the signal for the cell strands to separate were missing, the end product would have been a feather whose vanes were simple sheets of keratin. Such a feather would have been aerodynamically competent, if on the heavy side. Structures matching this general description have been described in a few Chinese specimens, such as the paravian *Epidexipteryx*, though just to confuse the issue, they were carried on its tail and were almost certainly used for display, not flight.

For the time being, we must accept a degree of mystery surrounding the origin of feathers. What we can say, however, is that they gave birds hitherto undreamed-of evolutionary options. Because feathers are inherently stiff, bird wings don't need to be actively tensioned, so their shape is nothing like as tightly constrained as in the other flying vertebrates. Bat wings all look pretty much the same, whereas bird wings can adopt all kinds of geometries, each suited to different locomotory ecologies, from the broad rectangles of vultures—great for soaring in tight thermals overland—to the long, tapering pennants of albatrosses, which give the exceptionally high lift-to-drag ratios that these huge seabirds need to stay aloft for many hours at a time riding gusts over the choppy Southern Ocean. The relative size of the wings is also freely variable. At one end of the extreme are the frigate birds that I mentioned in the introduction; at the other are penguins, whose miniaturized wings enable underwater flight, where the fluid density is eight hundred times

---

to get to their feeding grounds. Their fluffy down is a vital accessory that not only slows their descent but also cushions them from the impact of landing.

* These dinosaurs must have generated significant amounts of body heat for insulation to be worthwhile, but that's another story.

that of air. And there are all kinds of other customization options. Some owls maintain a frayed leading edge to their wings that lets them approach their prey in lethal silence. Hummingbird wings are extra stiff, such that they can be flipped upside down to get almost as much lift on the upstroke as on the downstroke; this, along with the birds' small size, allows hummingbirds to hover like insects. Because flight feathers overlap, the area of the wings can be altered at will, and if necessary collapsed completely, making possible the efficient bounding flight of songbirds, the plunge-dives of gannets and the precipitous stoop of peregrine falcons. The feather-enabled functional separation of wings and legs has allowed some birds to go flightless—something never seen in bats or, as far as we're aware, pterosaurs. The sprinting prowess of some of these former fliers—notably ostriches—now rivals that of cheetahs, just as the aquatic mastery of penguins rivals that of fish.

Birds are an essay in evolutionary serendipity. The group's astonishing locomotory success rests on several strokes of adaptive good fortune: the preadaptive tree-climbing facility of their dinosaurian forebears; the fact that their arboreal ancestors were in the right kind of forest to stumble on the selective pathway to gliding; the unique feather-based protowing design that not only opened the usually locked door to flapping flight, but also let them explore every possible facet of an aerial existence and more besides (burrowing is about the only kind of locomotion that no bird has mastered). Yet almost every step of this glorious evolutionary exploration was underlain by the same basic physical rules. More than any other one group, birds show just what the dance of natural selection and locomotory physics can achieve.

----------

We've now gained a pretty good impression of how evolution can squeeze the best locomotory performance—be that in the air, in the trees, or on the ground—out of a range of animal groups. But as you may have noticed, the majority of the creatures we've covered in these first three chapters have a few important characteristics in common—characteristics without which very little of what we've looked at so far could have happened. The template of two pairs of limbs, with a backbone holding everything together, has proved to be a rich source of locomotory options for natural selection to play with on land. But to see where these extraordinary structures came from, we are going to have to dive into the water, for the next stage of our evolutionary journey.

4

# Go-Faster Stripe

## In which we see how a knack for underwater wriggling took the vertebrates to the evolutionary front line

To bend like a reed in the wind—that is true strength.

—Lao-tzu, *Tao Te Ching*

In 348 B.C. an event took place that would forever change the course of biological thought: King Philip II of Macedon sacked the Greek city of Olynthus and sold its entire population into slavery. This may not appear to be an obvious addition to the timeline of science. Indeed, on the face of it, Philip's act was just one among many similar atrocities that pepper ancient history. However, the destruction of Olynthus had one seemingly insignificant consequence that would eventually lead to the publication of the *Origin of Species*. Prior to the attack, Aristotle had been living quietly in Athens for some twenty years, first as a student, then as a tutor at Plato's academy. Philip's actions made his position precarious in the extreme, for Olynthus was allied with Athens, and Aristotle was a Macedonian with close ties to the royal family (his father had been Philip's father's doctor). Afraid that he would suffer the same fate as Plato's mentor Socrates if he stayed, he quit the city* and

* Alternatively, his departure from Athens may have been precipitated by Plato's death, and the passing of the directorship of the academy to Plato's nephew.

made his way across the Aegean Sea to Atarneus, on the west coast of Asia Minor (modern Turkey), before founding a school at nearby Assos.

Aristotle's voyage, like Darwin's round-the-world trip on HMS *Beagle*, sowed the seed of a biological revolution. The journey took several weeks, during which time the fugitive philosopher would have been exposed to all kinds of wonders of the deep, courtesy of the fishermen with whom he traveled. He was surely staggered by the sheer variety of organisms that spilled onto the deck whenever a fresh catch was hauled up, but beyond the superficial riot of forms he also saw hints of an underlying order—that the creatures could be sorted into groups on the basis of certain shared characteristics. These days, of course, we know that such shared traits are indicative of common ancestry, but Aristotle, unlike some of his intellectual predecessors, didn't believe in transmutation: as far as he was concerned, the universe had always existed in its current state and would continue to do so forever. Indeed, this very idea of eternal stability may have been what the philosopher found so attractive about the natural world. Given the political turbulence of the times, who would blame him for seeking solace in a changeless, beautiful realm, against which the tumultuous affairs of mankind were but an insignificant blip?

The Aegean crossing was Aristotle's epiphany, and once safely settled in Assos, he resolved to do all he could to document the natural world and lay bare the rules by which it was organized. To that end, he set himself and his students to work observing, collecting, and dissecting the plants and animals of the area, all the while pumping local farmers, shepherds, and fishermen for tidbits of information, in an attempt to catalog and account for the distinguishing features of the different groups of organisms. Many of Aristotle's explanations seem strange to modern eyes, but this is hardly surprising, given that he had almost nothing to go on besides the abstract philosophy of the time, no tools except a dissecting kit, and no way of observing a living organism's internal workings. He thought, for instance, that the brain was a kind of refrigerator that balanced the heat of the heart, and that the projecting "nose" of a shark was an anatomical safeguard, deliberately making feeding difficult so that the voracious fish wouldn't die of overeating. Somewhat less understandable was his claim that men have more teeth than women. Every now and again, though, his thinking was spot-on: he correctly deduced the choking-prevention function of the epiglottis, for example, and figured out that hairs grow from their roots rather than their tips.

Although many of Aristotle's explanations of the form and function of organisms were wrong, his project had a deep and lasting impact. By his pioneering example he demonstrated that the natural world *can* be worked out, despite its dizzying diversity and complexity—that it has a rhyme and reason that might be comprehended by a sufficiently diligent and thoughtful observer. What's more, he showed how best to go about such an endeavor: to understand life, you must first classify it on its own terms. It would be well over two thousand years before Darwin revealed the true reason for the organization of the living world, but it was Aristotle who laid the groundwork, with a surprisingly accurate and long-lasting taxonomic scheme. Most of his categories of living things—such as mammals, crustaceans, cephalopods (octopuses, squid, and the like), cartilaginous fish (which rightly included rays), and whales (distinguished from fish if not then recognized as mammals)—are still in use.

Aristotle also had a clear sense of the nested, hierarchical arrangement of his animal groups, the higher ranks being distinguished by more fundamental features. He realized, for instance, that fish included both bony and cartilaginous varieties. At the highest level were two overarching categories into which all subsidiary groups could be sorted: the blooded (meaning the obviously red-blooded) and the bloodless (in reality, the not-so-obviously blooded). All blooded animals had sinews, a heart, a liver, a brain, and no more than four limbs or fins, whereas the bloodless generally seemed to lack those internal organs and could have more than four limbs. Furthermore, the blooded were marked by their possession of a backbone. This last attribute makes it obvious that Aristotle's categories are equivalent to the vertebrates and invertebrates, although these terms wouldn't be introduced until 1794, by Darwin's intellectual predecessor Jean-Baptiste de Monet de Lamarck.

These days, the idea of filing animals into either a vertebrate group or an invertebrate one smacks of anthropocentric chauvinism: the invertebrates are far more diverse than this scheme implies. But there's clearly something special about the vertebrates—something special enough to catch Aristotle's eye. As a group, we're not outrageously species-rich (the insects take that prize by an extremely wide margin), but our basic anatomical blueprint has proved so astonishingly adaptable that we're now major players in almost every ecosystem on Earth, from the deep sea to the skies. We've also managed to grow monstrously big from the perspective of the rest of the animal kingdom, giving us access to physical realms that few invertebrates can exploit. And with the exception of a few cephalopods, it's only among the vertebrates that high-level intelligence has evolved, and with it the ability to appreciate the rest of the world. What is it about this one group of animals that unlocked so many evolutionary doors? What's the secret of our success?

Answering this sort of question is often a difficult if not impossible task, because the outcomes of the evolutionary to-and-fro of life are dependent on so many factors, but I think a case can be made that one aspect of vertebrate biology has been of special importance in our rise to dominance. Although Aristotle spotted this key characteristic, it was Lamarck who gave it due emphasis by renaming our group in its honor. The characteristic in question is, of course, our spine. To understand the origin of our vertebral column is to understand how we were catapulted from wormy obscurity to the global big time. However, somewhat ironically, the primary selective benefit of this structure was its contribution to an activity that from the time of Aristotle to the end of the nineteenth century almost failed to attract any scientific attention at all. Where legs are (usually) for walking and wings for flying, the backbone is our answer to the physical problems of swimming.

## PUT SOME BACKBONE INTO IT

There is a famous thought experiment—one that I remember being introduced to at school—that considers what a person would look like if all the bones were removed from his or her body. I can still clearly recall the textbook illustration of the result—a somewhat disturbing, mandarin orange–shaped blob (a second picture of a person without any muscles was even worse—a pink puddle of skin surmounted by a head, with a jumble of bones pointing at all angles visible underneath). The take-home message from this little imaginative exercise was that holding our humanoid shape and moving around are absolutely dependent on our skeleton. Of course, we need only a passing awareness of the wider living world to see that this statement can't be true of everything, for there are plenty of familiar creatures, such as earthworms, that get by without any rigid material in their body at all.* But what's less obvious is how they do it. In Chapter 1 we saw that every muscle needs an antagonist to carry out postcontraction relengthening duties. Getting this to work is easy enough when rigid levers are available to make the contraction of one muscle lengthen another. How do worms manage without?

* Actually, earthworms do have short, stiff, retractable bristles, but they don't affect the overall mechanical challenge.

The answer to this question is straightforward if our definition of *skeleton* is broadened beyond the obvious. In a typical worm, such as an earthworm, the primary musculature is laid out in two concentric layers that usually extend nearly the entire length of the body. The outer layer, just beneath the skin, contains circular muscles, so named because their contractile fibers run parallel to the circumference of the body. The inner layer contains longitudinal muscles whose fibers run fore to aft, perpendicular to the fibers of the circulars. Internal to the muscle layers is a fluid-filled sealed cavity called the coelom (pronounced "SEE-lum"), from the Greek for "hollow," and inside that is the gut: a standard worm is essentially a tube within a tube. The coelom is a widespread and ancient animal feature (we vertebrates have one: it's the cavity between our guts and our body wall), and believe it or not, it functions as the worm's skeleton. When the circular muscles contract, the worm gets thinner, but this cannot proceed without consequence. Because it is completely sealed, the coelom's volume cannot decrease—not unless the forces produced by the worm's muscles are of superheroic magnitude. The diameter of the body can shrink only if the length of the body increases at the same time; when the body elongates, the longitudinal muscle fibers elongate, too. Conversely, when the longitudinal muscles contract, the worm can shorten only if it also gets fatter, and if its diameter increases, its circular muscle fibers extend. In other words, a worm's longitudinal and circular muscles antagonize each other, thanks to the pressure developed in the fluid of the coelom when one set contracts. It is this antagonistic action that moves worms around.† The technique—known as peristalsis—works particularly well when burrowing, but it can also be used to crawl on a surface. Lengthening the body with the circulars pushes the tip ahead, then the rest of the body catches up when the longitudinals take over.‡ The whole process is very easy to observe. Pick up an earthworm and you'll see the periodic elongation and bunching up of the segments caused by the alternating contractions of the circular and longitudinal muscles.

So much for worms. Vertebrates—even those with the simplest, wormy appearance, such as eels—do things entirely differently. An eel, like a worm,

† Not all worms have a coelom; some, such as flatworms, fill their body with loosely organized packing tissue (parenchyma) instead. But like the coelom, the parenchyma has a constant volume, and thus also converts a squeeze in one direction into an extension in another.

‡ For this to work properly, the worm needs to place anchors (the retractable bristles mentioned in the footnote on page 92) at certain moments of the cycle, or else it will just shorten and lengthen in place and go nowhere.

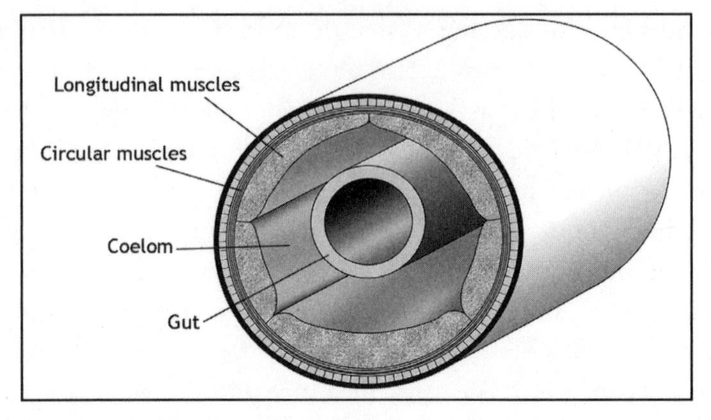


**4-1:** A diagrammatic section of a typical worm, showing the concentric layers of circular and longitudinal muscles, and the coelom within.

can be described as a tube within a tube, and its body wall is similarly packed with blocks of muscle from tip to tail. But these muscles are all longitudinal—no vertebrate has circular muscles in its body wall. We can get away without them because of our all-important vertebral column. Thanks to the rigidity of bone, the spine stops the body from shortening, so that if all the longitudinal muscles were to contract at the same time, nothing would happen, except perhaps a hernia. However, if only the right-hand set of longitudinal muscles contracts, the body bends to that side, because although the spine is inherently resistant to compression, its breakage into small segments—the vertebrae—gives it the required flexibility. That is, the inability of a vertebrate body to appreciably shorten means that a bend to one side lengthens the longitudinal muscles on the other. Here is the necessary antagonistic relationship.

There are many ways in which we vertebrates make use of the compression-resistant bendiness of our spine—indeed, we've seen a fair few in the last three chapters—but the standard method, as practiced by most fish (including eels) as well as lizards, snakes, and salamanders, involves the sequential activation of the muscle segments from tip to tail, with the wave of contraction on one side just ahead of that on the other. This activation pattern imparts an S-shaped wriggling undulation to the body. Now, one has only to glance at snakes to see that this method of movement works pretty well on a surface, but its widespread expression in fish immediately indicates that undulations are great for swimming. That's important, because worm-style peristalsis, which also works pretty well on a solid surface and wonderfully underneath it, is manifestly useless when it comes to swimming. This is why the origin of the bendy, compression-resistant vertebral column was so significant: any newfound access to a particularly effective and efficient means of transport, as occurred when birds developed wings, presents a wealth of evolutionary opportunities. We'll examine that crucial step later. First, however, we have a big question to answer. It's not difficult to see why wholeheartedly launching into the aquatic realm was such a big deal in evolutionary terms, but what was it about the undulatory movement pattern that opened this door? Why does wriggling the body from side to side lend itself so well to moving through the water?

## TREADING WATER

Aristotle was fascinated by animal movement: it forms a central theme in his great natural history treatise *The History of Animals*, and is the explicit subject of *The Motion of Animals* and *The Gaits of Animals*. Indeed, motion in the more general sense was one of the cornerstones of his philosophy—one which found application in his thinking about causality, ethics, and physiology. But in attempting to explain locomotion itself, he made a critical error. He reasoned that for movement to be initiated by an animal, "There [must] be without it something immovable . . . For were that something always to give way (as it does for mice walking in grain or persons walking in sand) advance would be impossible." For walking and running, this statement seems to make complete sense, but for swimming it just doesn't work. Water is anything but immovable, so aquatic locomotion should be as supposedly impossible for a fish as walking in grain for a mouse.

The critical piece missing from Aristotle's analysis was momentum. Our grasp of this enormously important concept developed gradually from the middle of the first millennium A.D., when Byzantine philosopher John Philoponus realized that there was a problem with the ancient movement paradigm. Aristotle's belief, introduced in Chapter I, that violent motions required the constant application of a force sat awkwardly with projectiles—if this were true, a thrown ball should drop to the ground as soon as it has left a person's hand. Now, Aristotle had thought of a way out of this: he reasoned that a ball in flight must leave a transient vacuum in its wake that's quickly refilled by an inrush of air (nature proverbially abhorring said vacuum). In his view, that selfsame air current would collide with the ball ahead and keep it going. However, Philoponus spotted that if air resistance is added to the mix, the air would have to both slow the ball down and keep it moving, which makes no sense. He reasoned that there must therefore be something intrinsic to a projectile that keeps it in motion. While it took some time before the physics was fully sorted out, we now recognize that something as momentum.

Momentum is mass × velocity (not speed—the direction is important, and only velocity contains this information), so a small object moving quickly has the same momentum as a large object moving slowly. Like energy, the momentum of a closed system (one that exchanges no matter with the outside world and isn't subject to external forces) is conserved. Should two objects with equal but opposite momentums collide and come to rest, their total momentum will obviously be zero, as they're no longer moving. But, less obviously, their total momentum is also zero prior to the collision, because the objects are moving in opposite directions—the positive momentum of one exactly balances the negative momentum of the other. The total remains unchanged.

This simple law of conservation is vitally important when trying to make sense of underwater locomotion. A fish swims by accelerating water backward—in other words, by imparting backward momentum to the water; because the total momentum of the system must be conserved, the fish itself gains forward momentum, and off it goes (note the resemblance to Newton's third law, which is essentially an expression of the law of the conservation of momentum in terms of forces). That's not to say that this is a swimming-only phenomenon; the law applies to all forms of locomotion. When you start walking, for instance, you impart backward momentum to the Earth. If you're facing east, this results in an imperceptible slowing of the Earth's rotation (or a speeding up if you go west). Because Earth is so massive relative to a human being, its change in velocity is minuscule, and in any case the momentum is returned when you stop (braking entails an imparting of forward momentum), so if you were planning on using this technique to shorten or lengthen the day, I'd think again. Flight can also be described in these terms: thrust involves the *rearward* discharge of momentum, lift a *downward* discharge, the forces being equal and opposite to the *rate* of transfer of momentum to the wake.

That's all well and good. But how does the vertebrate undulatory movement pattern bring about the required rearward acceleration of water? It took us an awfully long time to figure this out. The first person to attempt a rigorous explanation was Giovanni Borelli (1608–1679), a student of Galileo who later became an important Renaissance physiologist and physicist in his own

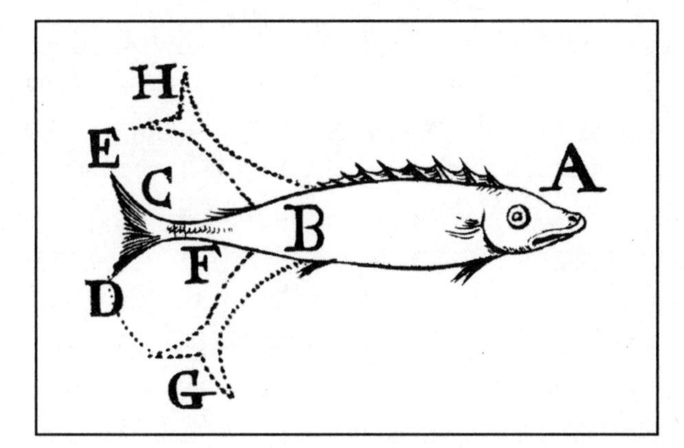


4-2: Borelli's reconstruction of a fish's swimming movements. From *De Motu Animalium* (1680).

right. Borelli had a lifelong interest in animal locomotion, particularly the underwater kind. He designed the world's first self-contained underwater breathing apparatus and the first submarine with a ballast system, but like Leonardo da Vinci's helicopter, these devices never left the drawing board. He believed that fish swim using their tail as an oar, with a power stroke and a recovery stroke. This accords well with our own experience of underwater locomotion. When we swim, our arms and hands directly push back on the water, giving it rearward momentum and propelling us forward. Or, to put this in more familiar terms, the backward movement of the arms during the power stroke generates a forward-pointing drag force, and as long as this isn't canceled out by the backward-pointing drag developed on the recovery stroke (when the arms have to be moved forward to reset), we get the needed thrust. Altering the trajectory of the stroke or the orientation of the arms during recovery can do the trick, as can lifting them clear of the water.

The drag-based, rowing model of fish locomotion held sway for about 250 years, but eventually the cracks in the theory began to show. The big problem was the fact that the back-and-forth beating of a fish's tail seemed to be symmetrical—it was impossible to reliably identify one direction as the power stroke and the other as the recovery stroke, although a few people managed to convince themselves that the beat to the left was a bit faster than the beat to the right, or vice versa. Scottish naturalist James Bell Pettigrew even added in 1874 that a fish must twist from side to side to feather the tail for its recovery stroke, just as an oar is twisted by a rower. This sort of thinking was finally crushed when our old friend Marey weighed in on the problem. As with his analyses of terrestrial locomotion, he felt that chronophotography was the best way to uncover a fish's secrets, despite the technical difficulties involved: after all, a swimming fish is much less constrained in space than a horse galloping down a racetrack. Marey's characteristically ingenious solution was to build a special elliptical aquarium—basically a fish racetrack—and attach an inclined mirror to the top, which allowed him to film a swimming fish from above. His photographic sequences of an eel—one of his first experimental subjects—immediately ruled out any notions of tail-twisting or some other kind of beat asymmetry: the beat of the tail to the left was identical to the beat to the right. That meant that the old drag-based model, with its power strokes and recovery strokes, could not apply. What Marey saw instead was that the undulatory wave of the eel's body moved backward slightly faster than the eel moved forward. That made its movement quite different from that of a slithering snake, despite outward appearances: a snake's body wave moves backward at the same speed as its body moves forward.

The relatively high speed of an eel's body wave makes it look as if the creature is constantly pushing back on the water. However, this is an illusion—the waves of an undulating fish may move backward, but any given segment of the body in fact moves from side to side. Following Marey's pioneering studies a great deal of work was done, both theoretical and experimental, to understand why such side-to-side movements cause a rearward acceleration of water. There's one particular project I'd like to mention, for it was carried out in my old stamping ground—the Department of Zoology at Cambridge University. Here in the 1950s Richard Bainbridge built what he called his fish wheel, a circular tank roughly 6 feet in diameter that looked rather like a giant transparent tire. This piece of apparatus, which acquired considerable notoriety at the time, was used as a kind of fish treadmill: it could be spun around by an electric motor at speeds of up to 20 miles per hour, so that a hapless swimming fish could be held in the same position relative to a static observer and photographed. Given that the entire whirling apparatus when full of water weighed about a quarter of a ton, this was not a task to be undertaken lightly!

## WAKEY-WAKEY

As wonderful as these early experiments were, a critical piece of information remained stubbornly out of reach. If you really want to know how a fish moves, you need to see what it does to the water, which is obviously tricky, water being completely transparent. The first attempt to get around this problem was made by American engineer Moe William Rosen in 1959. He came up with a way of making a fish's wake visible—by carefully adding water to a shallow layer of milk without causing any mixing. As his fish swam above the interface of the two liquids, its wake appeared as the milk layer was disturbed, in the form of a regular series of whirling eddies (or vortices, to give the technical name). Later workers added small particles to the water that achieved a similar effect without requiring such extraordinary skill (or I imagine, frustration). However, neither technique gave anything more than an impression of what was going on—the details remained elusive.

The impasse was finally broken by George Lauder, a professor of ichthyology (fish science) at Harvard. He borrowed a laser-based method from the field of engineering to both literally and figuratively illuminate a fish's wake. His lab, which I had the great pleasure of visiting in the early stages of writing this book, resembles nothing so much as a Rube Goldberg workshop—packed with equipment, computers and tools, with cables sprouting everywhere and ominous signs scattered about warning of the dangers of looking at the laser with your one remaining good eye. The laser is blue, so orange-tinted goggles are standard issue. At the heart of the lab, surrounded by high-speed video cameras, is a flow tank—the modern, straight (and safer) equivalent of Bainbridge's fish wheel. Most fish will happily keep swimming in the flow, and with practice can be held in position relative to the video cameras and laser by an adjustment of the flow speed. The wake-visualization method is essentially an extension of Rosen's milk-based technique. The flow tank is seeded with thousands of tiny plastic particles that make the wake visible. The laser's role is to make the movements of those particles intelligible. In one common version of the protocol, a sheet of laser light is aimed at the fish's wake and filmed in stereo. As the particles move through the laser sheet they become brightly illuminated and can be easily picked out by computer in the video footage. By keeping track of single particles from frame to frame, the wake (or at least the slice of wake that's illuminated) can be precisely reconstructed, enabling accurate measurements of the velocity of the fluid and the all-important momentum. This technique is called digital particle image velocimetry (DPIV). While a laser sheet only gives information about a single slice of the wake, it can be scanned through the tank to build a full 3-D picture, and recent advances allow the instantaneous tracking of particles in large volumes of water, giving us an even clearer picture.

Thanks to George Lauder's breakthrough, we're now much better informed about what fish do to the water. Figure 4-3 shows what's happening

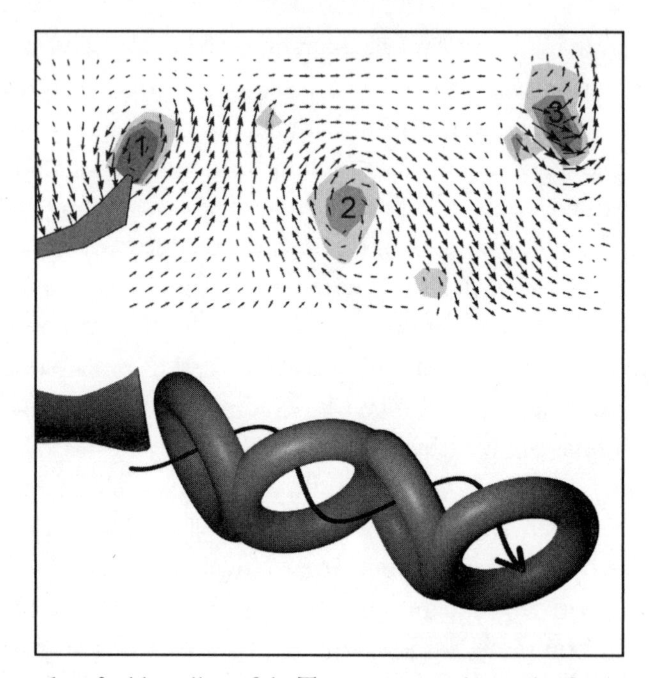


4-3: The wake of a bluegill sunfish. The top image shows the flow pattern in a horizontal slice of water behind the tail, as revealed by particle image velocimetry. The shaded regions are vortices, shed from the tail every time it reverses direction. The full 3-D reconstruction (bottom) shows that the vortex chain is actually a cross-section of linked vortex rings, through which snakes a jet of water.

behind one of Lauder's favorites, the bluegill sunfish. In cross-section, its wake takes the form of a staggered series of whirling vortices that are shed from the tail fin every time it reverses direction for a new stroke. Each vortex spins in the direction opposite to that of its immediate neighbors, so down the chain the direction of rotation alternates between clockwise and counterclockwise. That much was uncovered by Rosen. However, in the full 3-D rendering provided by the new technology, the alternating vortices are revealed to be a series of linked vortex rings, with a continuous backward jet of water snaking through the chicane in a smooth zigzag. The momentum of this jet is the source of the fish's thrust.

Bearing all that in mind, the wake behind a swimming eel (Fig. 4-4), reconstructed by Ulrike Müller and her colleagues at the University of Groningen, the Netherlands, may come as a surprise. The vortex rings are still there, but instead of linking together in a zigzag chain, they're flung out sideways, and the backward jet is nowhere to be seen. Eric Tytell, an old colleague of George Lauder's now running his own lab at Tufts University, Massachusetts,

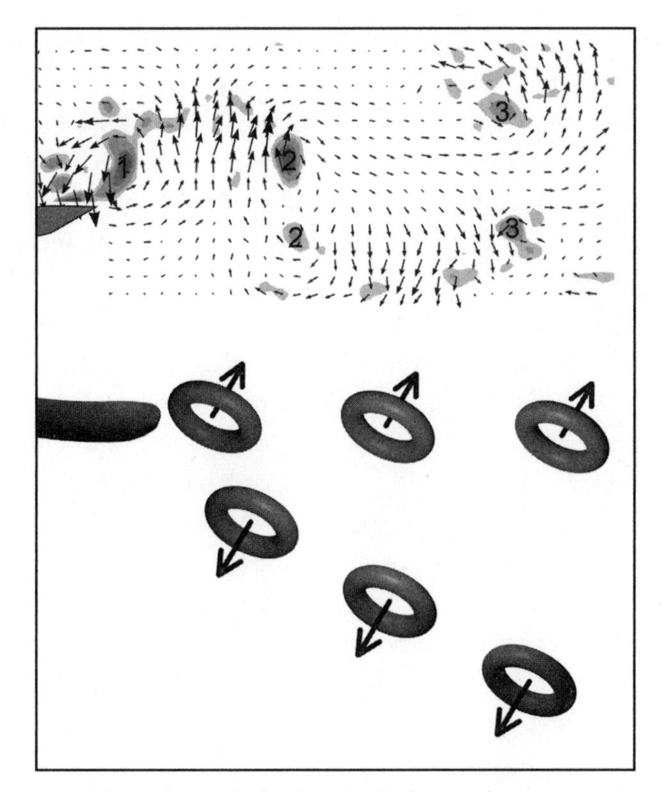


**4-4:** The wake of an eel in 2-D (top) and 3-D (bottom), showing the separated vortex rings and their lateral jets.

has an ingenious explanation for this apparent oddity. When swimming steadily, a fish need only impart enough momentum to the water to counteract its drag. An excess results in acceleration, a shortfall in braking. The idea that all forward propulsion must entail a backward jet seems more intuitive because we're used to boats, where all the thrust comes from a propeller or oars and the bulk of the drag from the hull. The propeller must therefore do more than its fair share of momentum-imparting work, for it has to counteract not just its own drag but that of the rest of the vessel as well. Sunfish (and other typically fish-shaped fish) are broadly similar: the tail does the lion's share of the thrust work, whereas the body is mostly responsible for the drag. Eels, and any other fish shaped like an eel, aren't like this—their entire body *is* the propeller—responsible for all the thrust and all the drag. When an eel swims steadily, there's therefore no reason to expect a jet in the wake—by this point the momentum budget has already been settled, for its undulating body has accelerated water backward by precisely the same amount as it's slowed down the relative backward movement of water near its body through drag.

All that's left is the sideways momentum, so the wake takes the form of lateral jets and their attendant vortex rings, gradually moving outward.*

All that's lovely of course, but our big question—how does undulating the body cause a rearward acceleration of water?—remains to be answered. But fear not, dear reader, for despite the fact that it's more difficult to analyze the flow near the body than the flow in the wake, Tytell and Lauder have done it. Once again, an eel was the experimental subject. The DPIV analysis showed that as the eel undulated its body from side to side, water was sucked into the troughs of the body waves (mainly from the neighboring crests) as they moved from head to tail. Because the body wave was moving in a rearward direction faster than the eel was moving forward, the water drawn into the troughs was accelerated not just sideways, but also backward. When a trough reached the tail, the accelerated mass of water was cast into the wake.

This use of negative pressures by an undulating eel to effect momentum transfer is strongly reminiscent of flight; indeed, this is technically a lift-based method of generating thrust (although the lift force vectors are in a horizontal plane because the body moves from side to side). It's a much more efficient way of producing thrust than a drag-based method, because in converting the wide sideways sweep of the tail end into a backward acceleration of water, a large mass is thrown into the wake. A drag-based method is necessarily more direct: only the relatively small volumes of water shifted by the backward-moving parts of the body can contribute to thrust. That may not seem like such a big deal, for the velocity can always be increased to compensate. However, because kinetic energy is equal to the mass of whatever's moving multiplied by the *square* of its velocity, divided by two—$1/2mv^2$—a given wake momentum can be achieved much more cheaply by accelerating a large mass of material up to a small velocity than a small mass to a large velocity. For example, a mass of 10 kg (about 22 pounds) moving at 2 meters (6.5 feet) per second has a kinetic energy of 20 joules (one joule is the energy it takes to lift a typical apple one meter [about 3.25 feet] off the ground), whereas a mass of 2 kg (4.5 pounds) moving at 10 meters (about 32.8 feet) per second, which has equal momentum, has a kinetic energy of 100 joules.† These energetic considerations explain why most fish have deep tail fins—the deeper the tail, the greater the mass of water accelerated backward for each sweep, and the cheaper the ride.

---

* When an eel accelerates, though, its wake looks a lot more like that of the sunfish.

† When the thing to which you're giving backward momentum is the size of a planet, as in walking, you barely have to expend any energy on this part of the process at all.

With its high propulsive efficiency, it's no wonder undulatory swimming has proved so popular among the aquatic vertebrates. That said, the method does have its drawbacks. For one, the high velocity of the water near the surface of the body compared to that experienced by a similarly shaped rigid body moving at the same overall speed adds substantially to the drag. And that's not all. One might suppose that each body wave acts as a kind of water syringe whose plunger starts at the head and moves toward the tail. However, as George Lauder's DPIV experiments have shown, the troughs of each body wave aren't compelled to take water only from the crest in front—each draws water *forward* from the crest behind as well. Only near the tail does the backward acceleration of water proceed unimpeded. That means that the undulations only become useful water accelerators at the back end of the body—elsewhere, they just cause a kind of back-and-forth sloshing. Why, then, do the undulations start so far forward in eels and the eel-like, especially given the aforementioned drag costs? My guess is that this is the easiest way to get a nice big sweep at the tail end—by gradually building up the wave amplitude from the front.

The relative hydrodynamic futility of undulating the front of the body hasn't gone unnoticed by natural selection. As the tail is doing most of the useful water pumping anyway, many fish confine the back-and-forth sweeps to their rear end. The ultrafast tuna are the most dedicated followers of this strategy. These have a very stiff body—the undulations are almost entirely confined to the tail, with the power being transmitted via tendons from the bulky musculature up front. The tail fin itself is deep but remarkably slender, with an extremely narrow base adorned with streamlined keels to minimize the drag associated with the side-to-side beating. Basically, tuna are absolute masters at imparting a whopping momentum to the wake with minimum cost. It's worth noting that these stripped-back undulations make the lift-based nature of the enterprise particularly obvious: a tuna essentially uses its long, thin tail to fly underwater, albeit with a vertical wing. Bearing that in mind, we can now see why flying animals with long, thin wings—such as albatrosses—enjoy such an easy ride. The great length of such wings relative to their area allows them to impart the necessary downward momentum associated with lift production very cheaply, by accelerating a large mass of air to a low velocity. The principle is identical.

Fish have one more propulsive trick at their disposal, and it concerns their vortex rings. Vortices form whenever a fast-moving body of fluid shears past a slower-moving body of fluid: as the jet scoots past its sluggish surroundings, friction drags some of the slower ambient fluid molecules with it. The pressure drop caused by that local depletion imposes a centripetal acceleration, and the fluid responds by rolling into a ring. Now, this vortex formation business, pretty though it is, might seem nothing more than an irrelevant quirk of fluid dynamics, but it turns out that fish take great care when making and shedding their vortices. Note that the formation process starts with the abduction of slower-moving fluid molecules by a jet. That ambient fluid winds up inside the vortex—if a dye is used to mark the ambient fluid, it can be seen that the ring is a veritable jelly roll of alternating fluid layers, some from the jet, others from the surroundings. A vortex ring is therefore a more massive structure than you might think, and as we saw earlier, a massive wake is a good wake. But the timing has to be just right. There comes a point when a growing vortex ring can't accept any more juice, whereupon it pinches off and drifts downstream. Any further discharged fluid forms a continuous splurge, lined with mini-eddies, which trails behind the departing ring. Such a flow is useless at dragging ambient fluid with it, so attempting to prolong the vortex ring formation process too far is an inherently inefficient way of generating thrust. Equally, however, a fish doesn't want to preempt the pinch-off and wind up with a meager runt of a vortex ring. To get the maximum benefit, a fish must hit the sweet spot precisely.

Remarkably, it seems that a whole range of organisms that use oscillating body parts to impart fluid momentum are on the same page as far as vortex ring formation is concerned, and I'm not just talking about swimmers. Animal flight expert Graham Taylor and his colleagues at the University of Oxford surveyed locomotory data from various swimming and flying vertebrates as well as flying insects, and found that the frequency at which they beat their wings or tails multiplied by the amplitude of the beat divided by their preferred velocity—a quantity called the Strouhal number—showed very little variation across a vast range of body sizes, regardless of whether the animals were moving in air or water. John Dabiri of the California Institute of Technology thinks that this extraordinary consistency is a result of the physical constraints on optimal vortex formation, for the key parameters that affect this process, such as formation time and the relative velocity of the vortex-feeding jet, are directly related to the parameters of the Strouhal number. The same principles have even been shown to apply inside our heart, as blood is squeezed with optimum efficiency from chamber to chamber.

## ACHING BACKS AND OLD FRIENDS

Now that we know how the undulatory movement pattern works, it's time to turn back to the thing that makes it all possible—the vertebral column. How did this wonderful apparatus evolve? This might sound like a formidably difficult question. The backbone has had a long, long history of evolutionary adjustments and upgrades, and the myriad versions currently in existence are bewilderingly complex structures. We're unlikely to come anywhere close to understanding the origin of the vertebrate hallmark by looking at such baroque gadgetry. However, nature has provided us with a convenient means of stripping away these layers of modifications. It often happens that animals recap certain aspects of their evolutionary history as they develop from a fertilized egg, and our vertebral column is one such example. Before the vertebrae begin to take shape, a strand of tissue forms along the embryo's long axis where the spine will be, just above the developing gut. It acts as a kind of master template, directing the formation of the spinal cord (not to be confused with the spinal column—the cord is the tube of nervous tissue running alongside it) as well as the muscles and the vertebrae. This structure is called the notochord, and it is our prototype spine.

As the vertebrae grow, they encroach onto the notochord's territory and eventually replace it. But not entirely. Remnants can still be found in our adult spinal column lurking *between* the vertebrae, forming the intervertebral disks. These are cushionlike packets filled with pressurized gel, and in mechanical terms are like the gel packs that you'll find at the more comfortable end of the bicycle saddle range: they even out compressive loads between two vertebrae and act as shock absorbers, reducing the risk of damage. Unfortunately, they're not themselves immune to damage. A particularly severe jolt can split a disk's outer jacket, causing the high-pressure contents to extrude. This is an infamous slipped disk (a misnomer, for the disk remains in place), which can be extremely painful should the bulging contents press on a nerve—a likely prospect, given the proximity of the spinal cord.

Unpleasant as they are, slipped disks tell us a great deal about our prevertebral spine, for as I've already hinted, in our distant evolutionary past, the notochord was our entire spine. But we can do better than that, because several primitive fish living today* retain the notochord in all its glory for their entire lives; these include the parasitic lampreys (a "surfeit" of which was apparently responsible for the death of King Henry I of England) and the delightful hagfish: eel-like denizens of the ocean depths that can tie

* I hesitate to use the word *primitive* for any species still with us. The retention of a purely notochordal spine is what's primitive in these fish; in other respects, they've undergone considerable modification since departing from the main branch of the vertebrate family tree. I mean no offense to the venerable creatures!

themselves in knots and, if bothered, will produce enough slime to fill a small bucket in a matter of seconds. Their notochords, rather like our intervertebral disks, are fibrous tubes filled with pressurized gel.

Significantly, notochords can also be found in two groups of *in*vertebrates. The first are the lancelets, a group of about twenty species of fish-like creatures, most less than 2 inches long, usually referred to as amphioxus (their defunct scientific name meaning "pointed at both ends"). Try to imagine what a very primitive vertebrate would look like, and you'll probably come close to amphioxus. It's a slender, transparent creature with rudimentary sense organs and no jaws, but it also has a spinal cord running along its back, a collection of closely spaced, gill-like slits that perforate its body at its front end (these are used to filter-feed), and a series of paired, V-shaped muscle blocks running along each side of its body, like a simple version of the musculature of fish. And it has a notochord, running from tip to tail just underneath the spinal cord—in exactly the same relative position as our spine.

The second notochord-bearing group of invertebrates seem about as far away from the vertebrates as it's possible to get. Sea squirts, visible at low tide along many of the world's coasts, are simple filter-feeding bags whose only obvious distinguishing features are two siphons: one for taking water in, one to get rid of it once the food has been strained out. They come in both solitary and colonial flavors, and are usually anchored to the spot, although a few enterprising forms—the salps—have converted themselves into jet engines by shifting their siphons to either end of the body, and now swim around with the plankton.* Only as larvae are sea squirts' vertebrate affinities apparent. These look like tiny tadpoles, with a tail containing a spinal cord, strips of muscle, and the all-important notochord. In the standard version of the life cycle, a larva swims around for a bit before diving to the seabed, where it glues itself in place and digests its tail—spinal cord, notochord, and all.

The presence of a notochord and the attendant segmented muscles in these two invertebrate groups is no accident. The sea squirts and lancelets are the vertebrates' closest invertebrate relatives, as was first pointed out by the Russian naturalist Alexander Kovalevsky in 1866, a mere seven years after the publication of the *Origin of Species*. Together, the three groups constitute the chordates, named after our trademark notochord.

---

* Salps can also go colonial. The extraordinary pyrosomes are glow-in-the-dark cylindrical aggregations of the creatures, open at one end so that the currents generated by the individuals combine into a single propulsive jet. Some species reach lengths of about 30 feet or more.

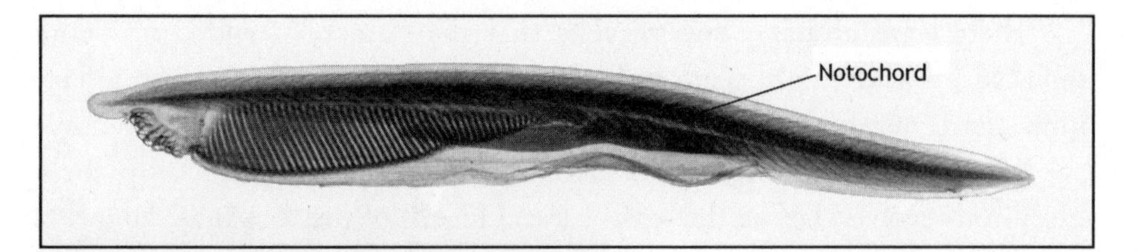

4-5: Two of the vertebrates' closest invertebrate cousins: the lancelet *Branchiostoma lanceolatum* (top) and the sea squirt *Ciona intestinalis* (left).

I imagine you're now curious about how such a manifestly fundamental characteristic works. Well, the notochord functions in the same way as the spine: it prevents shortening of the body, and thereby allows a chordate to bend its body from side to side. All the spineless chordates just mentioned can undulate like a typical vertebrate, and all lack circular muscles in their body walls. But how can this work, when the notochord lacks any rigid compression-resistant material? Remember that our intervertebral disks are pressurized—this is why they are at risk of rupture and why the contents extrude when a disk's jacket splits. The same is true of the notochord—the gel within the fibrous tube is under high pressure. As we all know through our experience with balloons and tires, high-pressure fluid can resist compression admirably well as long as the restraining skin doesn't burst. Indeed, this very resistance to compression exhibited by restrained fluids is what allows a worm's coelom to do its job, and in fact, the only mechanical difference between a typical worm and a notochord (aside from the internal pressure) is that the wall of the latter can't stretch in the circumferential direction: this is why the notochord can't shorten.†

† Roundworms, or nematodes, go a step further. Highly pressurized creatures, each with an inextensible skin, they are like isolated notochords (but with the musculature on the inside, not the outside). Unlike earthworms, they move by undulating.

There were probably two reasons that the notochord ended up being replaced by what might seem to be the functionally identical vertebral column. First, vertebrae can and do grow all kinds of extensions, which improve the leverage of the muscle blocks. If lampreys are anything to go by, these extensions evolved before the main central blocks of the vertebrae: lampreys have a fully notochordal spine, but with the addition of a collection of cartilaginous bits and pieces along its length that are almost certainly equivalent to our so-called neural arches, which are what you can feel if you run your hands down someone's back. Subsequently, the chunky bits of the vertebrae—the centra—were added to improve the existing compression-resistance function of the notochord, which enabled stronger contractions from the muscle blocks and more powerful swimming.*

## CORE STRENGTH

In the grand scheme of things, we can now see that the vertebral column is little more than a notochord plus, whose chief role is to beef up the function of the earlier pressure-dependent version. The real innovation was the notochord itself, and if we are to understand the evolutionary success of the vertebrates, it is the appearance of the notochord that needs explaining. A good place to start looking for clues might be the chordates' closest living relatives. Unfortunately, we're not going to find much useful information there, for we have some decidedly odd cousins. Most fall in the echinoderm clan: the headless and brainless starfish, brittle stars, sea urchins, sea lilies, and sea cucumbers with their unique five-rayed symmetry. We're not likely to find many points of comparison with the chordates among that group. The echinoderm fossil record is less perplexing—there are some extinct forms with an obvious head end and tail end at least—but there's still precious little information of relevance to chordate origins.

The closest relatives of the echinoderms—the hemichordates—are more promising. There are two basic models: the acorn worms (enteropneusts), which are long worms with gill slits and a vaguely butternut squash–shaped proboscis extending in front of their mouth, and the sessile, colonial

* Major differences in the method of growth of the vertebral centra in various vertebrate groups indicate that they evolved independently at least three times: in the sharks and rays, the bony fish, and the terrestrial vertebrates.

pterobranchs. As the name suggests, these animals were once considered our closest living relatives. In fact, marine biologist Walter Garstang (1868–1949) once proposed an elaborate theory for the origin of chordates that regarded the colonial pterobranchs as the starting point. He proposed that these sessile hemichordates gave rise to the similarly sessile sea squirts, from whose tadpolelike larvae the lancelets and vertebrates sprang by a process known as pedomorphosis. This happens when the timing of development is jolted such that an organism becomes sexually mature while still in its juvenile form. Put simply, Garstang's idea was that we vertebrates are sea squirts that never grew up. The scenario proved strangely irresistible at the time. Maybe there's something about the idea of turning our back on a tedious sessile life to fulfil our energetic, free-living destiny that appeals to our inner hero. Sadly, DNA evidence has put paid to this story once and for all. Hemichordates are closer to the echinoderms than either group is to the chordates; the pterobranchs are probably a specialized offshoot of the acorn worms; and worst of all, the sea squirts turn out to be more closely related to the vertebrates than are the more vertebrate-looking lancelets. They must therefore have become stripped-down filter-feeding bags only *after* they'd diverged from the rest of us. All told, there is no evidence of sessile ancestors lurking at the base of the chordate family tree whose shackles the vertebrates had to throw off.

So far, we've drawn blanks. But if our living relatives are no help, maybe we'll have more luck with the extinct ones. Vertebrates have an excellent fossil record thanks to our use of bone, but that's not going to help us much at the moment: to find out about the origin of the notochord, we have to go back to a time before the invention of our conveniently mineralized skeleton. Precious few known fossil deposits give us a detailed picture of such an ancient time. The Burgess Shale in British Columbia, discovered in 1909 by paleontologist Charles Walcott, is one. Dating back to the middle of the Cambrian period (505 million years ago), this famous formation has yielded an astonishing variety of animal fossils in an exceptional state of preservation: soft parts are visible as well as the more durable shells and skeletons. Two possible chordates or chordate relatives can be found there: *Pikaia* and *Metaspriggina*. Of the two, *Pikaia* is the more significant, because it appears to be the most primitive. It looks like a tiny-headed amphioxus, with typical segmented muscles, although the muscle block boundaries are shallower than the marked chevrons of amphioxus. It may also have had a notochord (ancient fossils are notoriously open to interpretation), but if so, it was much skinnier

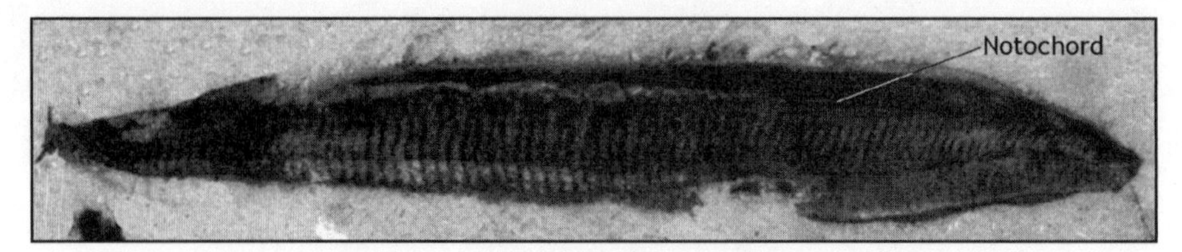


4-6: *Pikaia*, a probable chordate or chordate relative from the Burgess Shale.

than the modern version. *Pikaia* also had a unique elongated structure of unknown composition running along its back, which may have supplemented or even predated the compression-resistance function of the notochord.

For our current purposes, the evidently subsidiary role of *Pikaia*'s notochord is the most interesting thing about this beast. The presence of more or less typical chordate muscles in a body without an obviously functional notochord indicates that the undulatory style of locomotion probably evolved before the structure that we now associate with this form of movement. That might sound surprising, but if you think about it, there's nothing to stop a typical worm from indulging in a bit of undulation. All it needs to do is keep the body from shortening when the longitudinal muscles activate, which can be achieved with a bit of circular muscle tone: remember, the body can't get shorter if the diameter can't increase. Leeches, who are related to earthworms, are well known in biomechanical circles for this ability.

Suddenly, the mighty notochord seems to lose its sheen. But all is not lost. Undulatory worms differ from chordates in one critical mechanical respect: lacking a stiffened core, the longitudinal muscles of wriggling worms must work directly on the body wall to elicit bending, and require the mechanical support of the standard coelom to provide the necessary compression resistance. This contrasts sharply with chordates: in a fish, the contractions of the segmented muscles are transmitted to the notochord or vertebral column via the so-called myosepta—sheets of connective tissue that divide segment from segment. This is why the myosepta often have such a complex shape. If they were simple transverse sheets, the longitudinal muscle fibers would be appallingly inefficient at bending the spine. The oblique surfaces of the myosepta of modern chordates align more closely with the lines of action of their attached muscle fibers, allowing the muscles to pull more directly on the spine. What we see in *Pikaia* is an early move toward this unique centered take on undulations.

Why the chordates started down this road isn't clear. Thurston Lacalli, an amphioxus expert at the University of Victoria in Canada, has suggested

4-7: A sea bass fillet, showing the complex shape of a myoseptum.

that the primary function of the notochord was to stiffen the gut to suppress any unwanted effects of body undulations on gut motility. While a somewhat lowly starting point for a revolution, Lacalli's theory makes developmental sense, for in amphioxus the notochord arises as an outpocketing of the embryonic gut. But however it happened, the appearance of the notochord gave the chordates a fantastic new way of pulling an old locomotory trick. The leech method of undulation works just fine, but it doesn't tolerate any adjustments to the wormy body shape—the body wall must remain thin enough for the longitudinal muscles to do the bending work. Once the power transmission was directed to a stiffened core—our go-faster stripe—this constraint was wholly lifted. Chordates could thenceforth do whatever they liked with their body, within reason, without interfering with the undulatory essentials.

One need have only the vaguest of passing acquaintances with the chordates to see that they didn't look their notochordal gift horse in the mouth. Their adaptive radiation has been nothing short of breathtaking, with layer upon layer of transformations taking them on all kinds of morphological adventures. They were free to add powerful jaws or develop a streamlined, torpedo-shaped profile to minimize drag. Conversely, their body could become flattened in connection with a bottom-dwelling existence, as happened independently in the rays and flatfish. As we saw earlier, vertebrae were added to increase muscle leverage and augment the notochord's compression-resistance function, allowing larger size and more propulsive power. A deep tail fin could develop to increase thrust—a trend taken to an extreme in ambush predators, such as pike, whose rearward concentration of fin area gives whopping accelerations. More subtly, the shape and size of the vertebrae could be tweaked to vary the stiffness of the backbone down its length, and the power transmission could be altered by bundling the myosepta into tendons, so that the powerful muscles of the anterior body could remotely operate the tail—both opportunities were exploited to the full by tuna. Paired fins could be added for better locomotory control, which in some lineages have usurped many of the more routine propulsive duties from the body and tail, enabling yet more morphological flights of fancy that have given us such wonders as seahorses and ocean sunfish. Ironically, a worm shape has commonly been reacquired: eel-like forms have evolved repeatedly from more streamlined ancestors. While such a shape can't provide the powerful thrusts of a more typical fish body, eels enjoy a much more economical mode of locomotion at their leisurely pace.

----------

Thanks to their go-faster stripe, the vertebrates have become the undisputed masters of underwater propulsion, better suited than any other group to an active aquatic existence. But of the many wondrous twists and turns of the group's evolutionary diversification, there is one transition that ranks above all others for its sheer unexpectedness. Despite their success beneath the waves, the vertebrates found a way to colonize the land. In the next chapter, we'll look at this wholly improbable episode in the long history of locomotion.

# 5

# The Improbable Invasion

## In which we find out how our fishy ancestors took a deep breath and crawled out of the water

Let me, if not by birth, have lands by wit:
All with me's meet that I can fashion fit.

—William Shakespeare, *King Lear*

*Cute* is not an adjective commonly applied to sharks. *Vicious*, *deadly*, *menacing*, and assorted synonyms all spring easily to mind, thanks in no small measure to the movie *Jaws* and its celluloid ilk (and the more excitable brand of nature documentary). *Sleek* and *graceful* may also be used once in a while by those of us who think sharks have had more than their fair share of bad press over the years. But unlikely as the prospect may seem, I defy you to watch an epaulette shark go about its business and not feel your heart soften. This delightful species, named for the dark patches above the pectoral fins (the front pair), can be found in shallow coastal waters off New Guinea and northern Australia. They're reassuringly small—rarely exceeding 3 feet in length—and are dangerous only to the miniature crustaceans, worms, and fish that constitute their preferred prey. Their most endearing quality, however, is their method of locomotion. While they *can* swim, they generally do so only in an emergency. At other times, should they need to get from A to B, they go for a stroll over the seabed.

If and when you come across an epaulette shark, it won't take you long to realize that it's never going to win any speed records, for good reason. Lacking legs, the fish must instead impart momentum to the substrate using its fins, which seem little modified from the standard shark appendage. Imagine replacing your legs with Asian fans and you begin to see what our shark friends are up against. That they can make any headway at all is largely down to a nifty locomotory trick that enables them to extend the stride length to a distance well beyond that which their stubby fins alone could deliver. In fact, very little forward propulsion comes from direct retraction of the fins. These really only act as points of contact (or *points d'appui* as they are quaintly known in the field): the body is responsible for most of the forward motion, bending in a wide arc about a stance fin to advance the opposite swing fin. The power for this movement comes mainly from the body wall muscles that we looked at in the last chapter. When the muscles on one side of the body contract, the tips of the pectoral and pelvic (rear) fins on that side are brought closer together, while their opposite numbers move apart. As long as this happens while the pectoral fin on the contracting side and the pelvic fin on the extending side are firmly planted, forward movement will ensue. The resulting gait is a walking version of a trot. It sounds complicated, but all it requires is an ability to pass staggered waves of muscular contraction down each side of the body. As we saw in the last chapter, this should be no big deal for a shark: that undulatory style of movement is exactly what we vertebrates were made for.

So, it would seem that the epaulette shark has been effectively preadapted for its seabed-hugging existence, and as a result can forage among the coral reefs of its Australasian home with minimal effort. Which is very nice, but it hardly ranks as one of life's major evolutionary transitions. You may wonder,

5-1: An epaulette shark (*Hemiscyllium ocellatum*), going for a stroll.

then, why we're peering so closely at this unobtrusive creature. Well, unobtrusive it may be, but in seamlessly turning an ostensibly swimming mode of locomotion into walking, it's given us a valuable glimpse into one of the most extraordinary episodes in our long locomotory history: when the vertebrates—a group dedicated like no other to the life aquatic—upped sticks and moved onto land.

## RECURRING DREAMS OF DRY LAND

It might now seem that we have the answer to this apparent paradox. In many ways, including the literal, it's but a small step from walking around on a seabed, riverbed, or lakebed to doing the same on land. A solid substrate is a solid substrate, regardless of the state of the fluid sitting on top of it. However, any would-be landlubber faces a major obstacle. Because the density of air is eight hundred times less than that of water, any emerging animal instantly puts on an enormous amount of apparent weight. Holding the body off the ground becomes a test of strength (for muscles and skeleton), and any animal that would rather give up and drag itself along the ground must instead contend with large friction forces.

Bearing this in mind, we can see why the epaulette shark does all its walking on the seabed: like all sharks and rays, this species has an internal skeleton of cartilage—the relatively flexible substance that coats the ends of our long bones and that gives shape to our ears and nose. It serves perfectly well underwater, but it's not up to the task of supporting the body on land, so unless something truly extraordinary were to happen, sharks and their descendants will almost certainly remain water-bound for as long as they last. Bony fish, however, have proved less constrained than their cartilaginous cousins, thanks to their more robust internal chassis. Mudskippers—a group of specialized intertidal gobies—are probably the most famous amphibious fish, but there are many other examples, from walking catfish to various kinds of eel. Between them they've hit upon several different ways of moving around on land. Eels, for instance, being practically finless, move like snakes. Walking catfish also undulate from side to side (in a rather violent fashion), but use their pectoral fins as props to extend the stride length—rather like an epaulette shark that's lost the use of its pelvic fins. Some smaller fish are adept at jumping, using their tails to catapult themselves into the air. Most of these species move on land only occasionally, as a means of migrating from pool to pool or river to river. Mudskippers, being intertidal, are more regular users of the terrestrial environment. Their walking style is very different: it doesn't involve body undulations at all, and is instead driven by symmetrical thrusts from the pectoral fins.

A notable absence in the bony fish terrestrial movement repertoire is the trot performed by epaulette sharks. Except, that statement is not quite true: technically, a group of bony fish *did* settle on this walking style, but we don't call them fish anymore. They are the tetrapods: *tetra* in recognition of their use of all four paired fins on land, and *pod* because those fins have become so modified for their new purpose that they are no longer fins at all, but limbs. This is, of course, the group of vertebrates to which we ourselves belong, along with the other mammals, birds, reptiles, and amphibians.

There was clearly something unique about the way in which tetrapods became terrestrial. Even mudskippers—the most dedicated terrestrial non-tetrapod vertebrates—are unmistakably fish, and so are confined to a pretty narrow way of life, albeit a productive one. The tetrapods, on the other hand, left most of their fishy trappings behind long ago, and have radiated to fill every conceivable terrestrial niche that could have been filled by something as big as a vertebrate. To add injury to insult, they've repeatedly and successfully reinvaded the water, in many cases beating their distant fishy cousins at their own game. The colonization of land by the tetrapods must then surely rank as one of the most significant events in the history of life on Earth. By comparison, the terrestrial exploits of the other fish look halfhearted—mere footnotes in the ledger of evolution. That there must be something special about limbs—the tetrapod hallmark—is blatantly obvious. What's less obvious is why, despite the multitude of vertebrate landfalls, the conversion of fins into the more terrestrially competent limbs should have happened only once. The mudskippers must be kicking themselves. Well, they would, except . . . you get the idea.

## THE EMPEROR'S NEW FISH

Perhaps we shouldn't be too hard on the semiterrestrial fish. Fins and limbs are very different structures after all—so different that it was a long time before anyone spotted that the two kinds of extremities were at all comparable. While Aristotle was aware that nearly all vertebrates ("blooded" animals), be they terrestrial or aquatic, were provided with two pairs of appendages, he ascribed this similarity to the fact that being blooded, they had to move with four points of motion: only the bloodless (the invertebrates) use more than four. It would take the discovery of an extraordinary species of fish in extraordinary circumstances to go beyond such arbitrary pronouncements.

In August 1797 Napoleon Bonaparte, fresh from his victorious military campaign in Italy, wrote to the executive council of the Republic outlining an audacious proposal to invade Egypt. It was an odd request, but the council was keen to get the ambitious and popular young general as far from Paris as possible.* So, they agreed, and on May 19, 1798, Napoleon set sail from Toulon with a fifty-thousand-strong invasion force. But this was no ordinary military endeavor. Alongside the warriors of Napoleon's expeditionary force were 167 hand-picked scholars—mathematicians, historians, natural historians, physicists, astronomers, and more—tasked with studying every aspect of the exotic land they were about to invade. Among them was a twenty-six-year-old zoologist by the name of Étienne Geoffroy Saint-Hilaire.

Geoffroy (as he is usually known) is one of the great unsung heroes of evolutionary biology. While Darwin was chiefly concerned with how organisms change, Geoffroy was instead drawn to the idea of the unity of form: in the many shared characteristics between organisms, he saw evidence that there was a single ideal organism-type, or archetype, from which all species had been variously derived. He was thus an early proponent of what came to be called homology, a term coined in 1843 by British anatomist (and anti-evolutionist) Richard Owen, meaning "the same organ in different animals under every variety of form and function." To give the classic example, the wing of a bat, the arm of a human, the flipper of a seal, and the paw of a cat, despite their superficial differences, are clearly built along the same lines: all have a single bone at the base (the humerus) that articulates with the shoulder girdle, followed by two bones (the radius and ulna) that articulate with the first at the elbow joint, a collection of small wrist bones, and finally a series of up to five (barring polydactyl mutants) variously shaped digits. In the light of evolution, such homologous traits in different species are the essential evidence of their shared ancestry.

Geoffroy was arguably the most ambitious homology seeker that has ever lived: indeed, more often than not he was too ambitious. In his later writings he tried to bridge the seemingly unbridgeable divide between vertebrates and invertebrates by proposing that the exoskeleton of an arthropod was essentially the same thing as the endoskeleton of a vertebrate, the only difference being

* Not that it did them much good: Napoleon came back in 1799 and, following a coup d'état, installed himself as first consul.

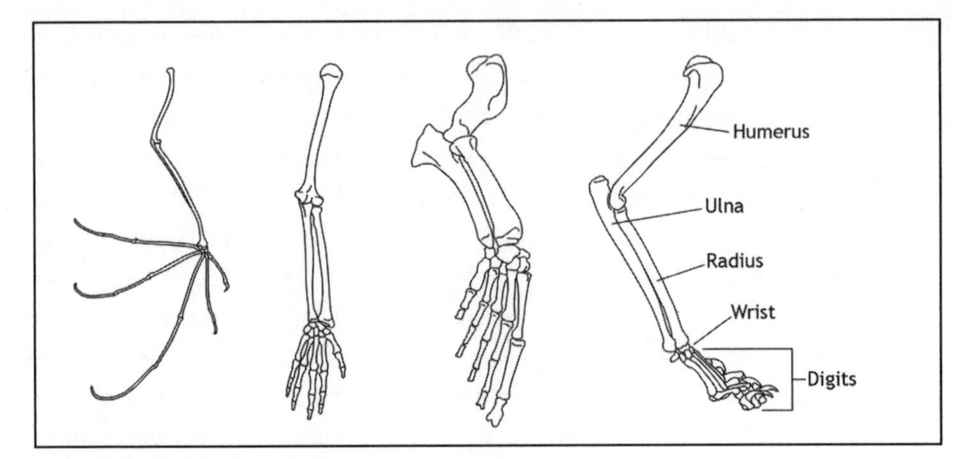


**5-2:** The forelimb skeletons of (from left to right) a bat, a human, a fur seal, and a cat. Despite the great differences in overall appearance, the homologous humerus, radius, ulna, wrist bones, and digits are clearly identifiable.

that arthropods lived *inside* their backbone. He also regarded cephalopods (squid, octopuses, cuttlefish, and the like) as vertebrates that had somehow folded backward over themselves. These extreme attempts at homologizing were rightly considered fanciful by contemporary anatomists, although recent genetic evidence suggests that another of his propositions—that vertebrate organization can be obtained from that of an arthropod or worm by flipping it upside down—may have some truth in it (more on this in Chapter 6).

Back in 1798, such thoughts were far from the young Geoffroy's mind as Napoleon's army successfully invaded Egypt. Geoffroy soon got to work describing the zoological wonders of his new home, and before long had enlisted his own small army of local fishermen and huntsmen to help him with his task. One can only imagine his joy as species after species, many of them unknown in Europe, were added to his collection. But it wasn't to last. On August 1, a mere month after Napoleon made landfall, Lord Nelson's fleet finally caught up with its quarry, and following a tactical blunder by the French commander, managed to destroy or capture all but four of the French ships. Napoleon continued to wage a land war in the Levant for a time, but a catalog of further woes from local uprisings to the bubonic plague eventually made his position untenable, and in 1799 he abandoned the campaign, slipping back across the Mediterranean to France. Two years later, the remaining French—including Geoffroy—surrendered, and were taken home on British ships. Under the terms of the capitulation, everything collected by the French scholars throughout their three-year incumbency, including the

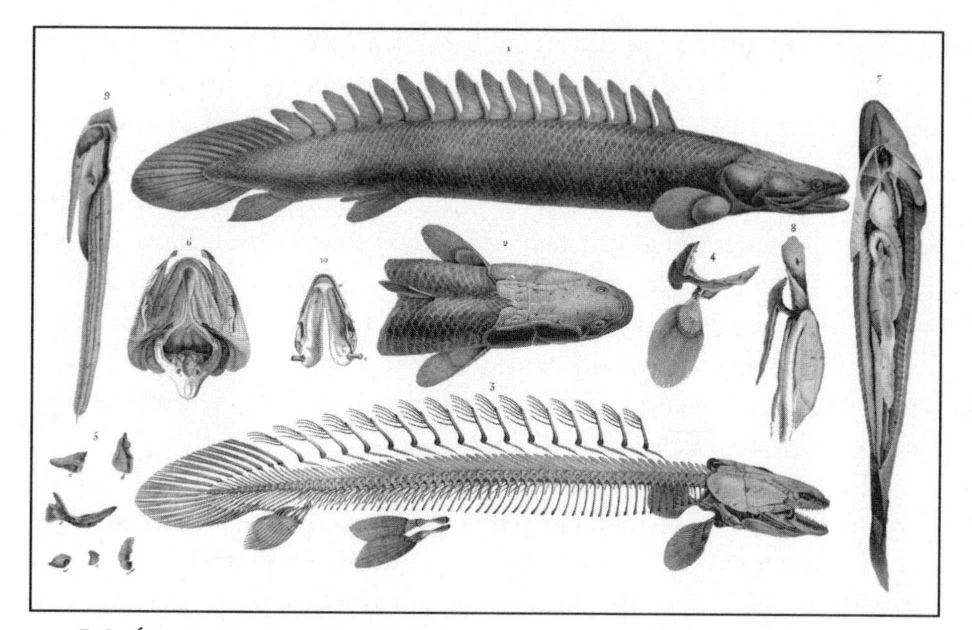

5-3: Étienne Geoffroy Saint-Hilaire's illustrations of *Polypterus bichir.*

fabled Rosetta stone, was supposed to have been confiscated. However, in a last show of defiance, Geoffroy threatened to set fire to his zoological collection rather than hand it over, so he was allowed to take it back to Paris (the Rosetta stone, however, now has pride of place in the British Museum).

Among Geoffroy's rescued specimens was a strange species of fish from the Nile. In 1809 he declared that if this had been the only creature he had found, it alone would have compensated him for all the terrors and hardships of the ill-fated campaign. It was about 20 inches long, with a narrow, cylindrical body covered in heavy scales, and a series of small dorsal fins extending along its back, after which Geoffroy decided to name it: *Polyptère* (many fins), later Latinized to *Polypterus.* Aside from some unusual features of its gills, two other characteristics stood out. First, the fish had a pair of air-filled sacs connected to its gullet, which Geoffroy interpreted as a swim bladder. Such an organ, used to regulate buoyancy, is commonly found in bony fish, although it usually takes the form of a single bag, not a pair. But the pectoral fins with their fleshy, lobed bases were what really caught the young zoologist's attention. In his own (translated) words, the fins were mounted on "genuine arms, since one can count inside them the same small bones as are [found] in mammals."*

* It has since been found that *Polypterus* can use its lobed fins, in conjunction with body undulations, to move around on land, rather like a slightly more elegant version of a walking catfish.

This was a remarkable deduction. Several years before Darwin was even born, Geoffroy noticed that a fish fin was comparable to a mammalian limb despite all appearances to the contrary, and although it would take many years for the deeper significance of his observations to be appreciated, his presentation of *Polypterus* to the world gave us our first hint that fins and limbs perhaps weren't so different after all. It soon became clear that this jewel of the Nile wasn't alone: the nineteenth century saw the discovery of a great many fossils of extinct fish whose fins, like those of *Polypterus*, were lobed. The group was formally named the Crossopterygii in 1861 by Thomas Henry Huxley (the name means "fringe fin," an allusion to the fringe of slender bony rays around the lobe).

The best known of the extinct crossopterygians is the Late Devonian pikelike *Eusthenopteron*, found in 1881 in the Miguasha Formation near Quebec. It will serve to illustrate what's going on inside a typical crossopterygian fin lobe. At the base of the fin skeleton, a single bone articulates with the pectoral or pelvic girdle, in contrast to the handful of miniature bones that support the fins of most bony fish. Beyond lie a pair of bones, the shorter of which supports another pair, and the pattern is repeated once more. The bones mark the extent of the lobe: beyond them lie the fringing rays of the fin proper. Significantly, that one-plus-two set of bones at the base is exactly what you find in a tetrapod limb: the humerus, radius (in front), and ulna (behind) in the forelimbs; and the femur, tibia, and fibula in the hindlimbs. The homologous arrangement is firm evidence that the ancestors of the tetrapods are to be found among the crossopterygians.

Since being described in *Eusthenopteron*, the same kind of fin skeleton, with a single bone at the base, has been found in many lobe-finned fish—enough to tell us that we're probably looking at the basic pattern for the entire group. Ironically, the odd one out is dear old *Polypterus*. The skeleton sitting inside the lobes of its fins is so different from *Eusthenopteron*'s—with two widely divergent bones articulating with the girdle and a broad fan of cartilage filling the space between them—that *Polypterus* was eventually ostracized from the crossopterygians despite being, in a sense, the group's founding member.

With the loss of *Polypterus* from the crossopterygian fold, and the dawning realization that tetrapods are their descendants, it was decided in the 1950s that a renaming was in order. We (for tetrapods are included) are now known as the Sarcopterygii (sarcopts for short), meaning "flesh fins," in reference to the lobes. Almost all the fishy sarcopts are long gone, although there

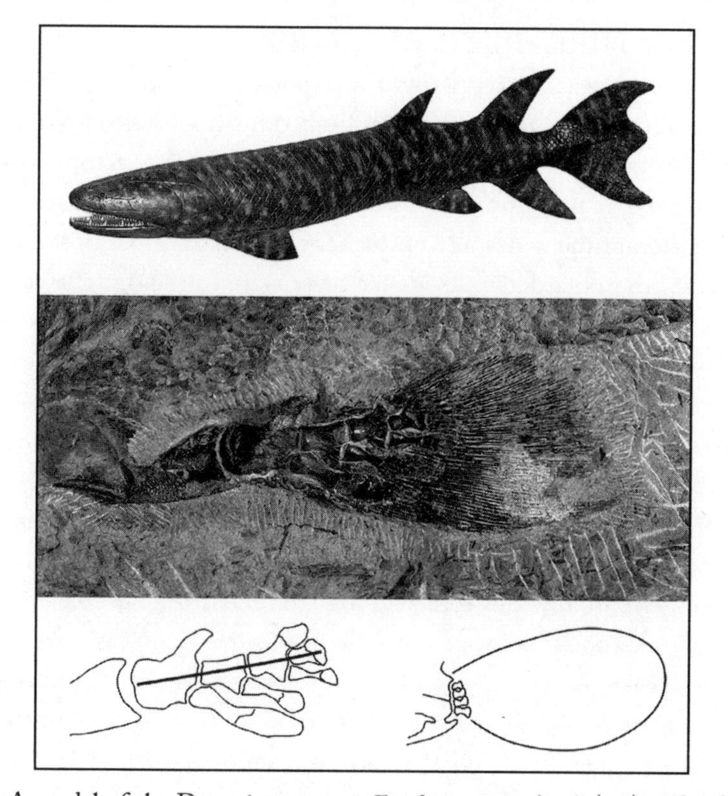

5-4: A model of the Devonian sarcopt *Eusthenopteron foordi* (top) and a fossil of its pectoral fin (middle). The back edge of the fin is toward the top of the image. The main axis of the fin's endoskeleton is indicated in the schematic below, alongside a diagram of a typical actinopt fin (note the extent of the fin fold), with a highly reduced endoskeleton and multiple bones articulating with the girdle.

are a few hangers-on: the freshwater air-breathing lungfish of South America, Africa, and Australia, and another group whose introduction I'll save for the end of the chapter. All other bony fish—perch, cod, goldfish, eels, mudskippers, and the like—are known as the Actinopterygii (actinopts for short), meaning "ray fins," on account of their general though not quite universal lack of fleshy lobes: the lobe-finned defector *Polypterus*, you see, is the most primitive living member of the actinopts. While a little confusing, this means only that the last common ancestor of the sarcopts and actinopts must have had lobed fins, and that the lobes were subsequently reduced during actinopt evolution. It's the single bone at the base of each fin skeleton, as seen in *Eusthenopteron*, which sets the sarcopts apart. All actinopt fins, be they lobed or not, have multiple copies of such elements.

## THE PREMATURE BIRTH OF LIMBS

With the link between sarcopts and tetrapods established, attention could turn to the nuts and bolts of the fin-to-limb transition. Paleontologist Alfred Sherwood Romer was one of the first to come up with a complete scenario, building on earlier work by geologist Joseph Barrell. Barrell had previously drawn attention to the widespread presence of red sandstones in the Devonian period, which he regarded as evidence of seasonal drought: the red color is caused by iron oxides, which were presumed to have formed when the sands were exposed to the air. As sarcopt fossils had been found in these Old Red Sandstones, Romer reasoned that the fish would have been exposed to the same drought conditions, and proposed that they used their lobed fins to trudge overland to waters new whenever their aquatic abodes dried out. Over time, so the story goes, the fins gradually became more robust—better suited for terrestrial use—and the relevant sarcopts gradually spent less and less time in the water. Eventually they returned only to breed, and became amphibians.

Romer's idea made a lot of sense: as we know, many fish are known to undertake occasional terrestrial excursions even without the help of lobed fins. Surely the sarcopts would have had an easier time of it, with their more limblike appendages. It's a good hypothesis, and like all good hypotheses, it's falsifiable. The central idea is that the fin-to-limb transition happened in a terrestrial setting. If a fossil more primitive than known amphibians were to be found with predominantly aquatic adaptations, but bearing limbs rather than fins, the whole fish-out-of-water scenario would be seriously challenged. Unfortunately for Romer, in 1987 just such a fossil came to light. The specimen in question was found, rather improbably, in the mountains of East Greenland by Jennifer Clack, curator of vertebrates at the University Museum of Zoology, Cambridge. She'd been tipped off by an earlier geological survey that tetrapod fossils were common in the area, so had managed to talk her way onto a Danish oil-hunting expedition to try her luck. To cut a long story short, her luck was in: among the many specimens sent back from the expedition was a perfectly preserved, nearly complete skeleton of a creature called *Acanthostega* that had until then been known only from its skull. When it was alive, its corner of Greenland had been a steamy equatorial river basin. That was about 365 million years ago, in the Late Devonian, making *Acanthostega* one of the oldest tetrapods known—certainly the oldest for which we had anything more than fragments of fossilized bone.

Back in Cambridge, as the skeleton was gradually freed from its stony sarcophagus, it became clear that *Acanthostega* was no ordinary tetrapod.

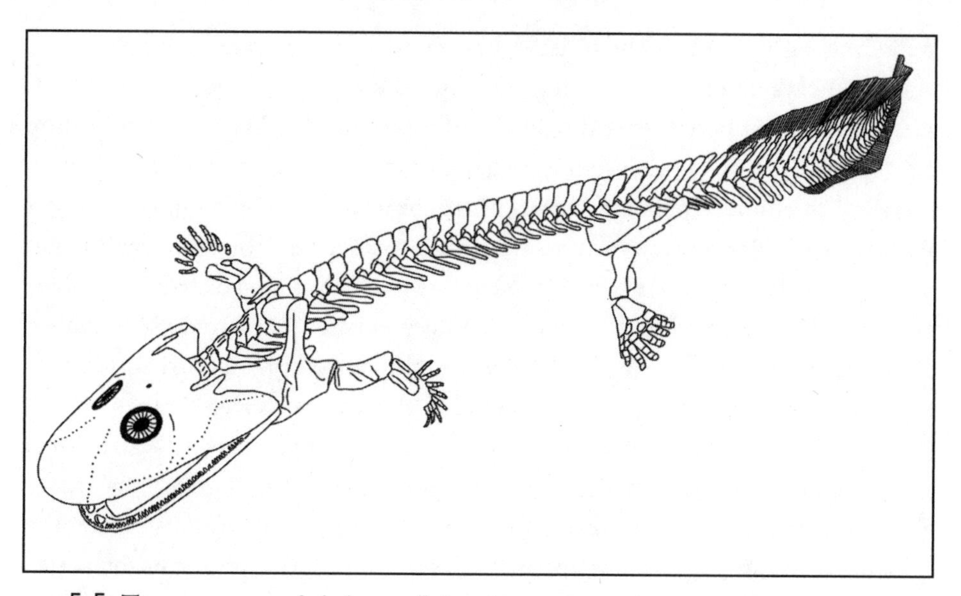

5-5: The reconstructed skeleton of *Acanthostega gunnari*.

While it looked superficially salamander-like, it had some decidedly fishy characteristics. Its tail carried a broad fringe of fin rays, and preserved within its throat was the delicate bony scaffolding of what can only have been a set of gills. And although it had limbs, they weren't the legs of a land walker. For a start, the hands and feet had eight digits apiece, whereas all other tetrapods then known had or have no more than five digits per limb. There were also no proper elbow, knee, ankle, or wrist joints—only vague regions of flexibility—and the shoulder was too restricted in its range of movement to enable a full walking power stroke.

Everything about *Acanthostega* cried aloud that it was an aquatic animal. Its ribbonlike tail was undoubtedly used for underwater propulsion, and its simple arms and legs, with their oversize hands and feet, wouldn't have been able to both prop the body up off the ground and push it forward at the same time. They were paddles, unsuitable for walking on land. These facts in themselves would not be enough to overturn Romer's scenario, for one could argue that, like many modern salamanders, *Acanthostega* was secondarily aquatic—descended from terrestrial ancestors that had previously converted their fins into limbs on land. However, this suggestion is inconsistent with three aspects of *Acanthostega*'s anatomy. For one, the presence of fin rays around the tail is not seen in any secondarily aquatic animal—even those that, like newts, have evolved flanges of skin to broaden the tail. Then there are the extra digits—also not seen in any tetrapod that has returned to the water. The real clincher, though, is the gills. While many larval amphibians have them, they're always external, branching out from the throat like fleshy pink feathers. *Acanthostega* had a typical fishy set of internal gills. We can thus be sure that its ancestors were every bit as aquatic as *Acanthostega* itself. Furthermore, contrary to Romer's beliefs, they had little need to be otherwise, for drought is not actually necessary for red sandstone formation. The iron oxides that give the rock its red color need not have formed in situ. Just as happens today, they may have formed on land and then been washed into streams, estuaries, lakes, and lagoons. Over millions of years, the red sediments turned to stone while fish (and the odd tetrapod) teemed in the waters above.

The reconstruction of *Acanthostega* as a fully and primitively aquatic animal immediately raises a tricky question: why were the paired fins transformed into limbs if terrestrial locomotion had nothing to do with it? The trickiness, however, is more apparent than real—borne of a kind of vertebrate chauvinism, which presumes that legs are only good on land, and that any aquatic creature worth its salt has fins instead. We have only to look at crustaceans to realize how unjustified this assumption truly is. Lobsters and crayfish are as fully aquatic as any fish, yet they get around on foot. Even ostensibly committed swimmers, such as prawns, have a full set of legs as well as the paddles that line their abdomen. The common factor here is substrate contact: regardless of whether you're underwater or on land, sturdy legs are generally more suitable than flexible fins where walking is concerned. The reason is obvious: when using an appendage-driven push on the substrate to move forward, you're going to want as much of the backward pushing force as possible to be delivered *to* the substrate, so that the all-important propulsive ground reaction force is nice and large. You don't want that force to be wasted on bending the appendage into a shape that's more likely to slip over the surface than push onto it. As far as a vertebrate's concerned, that means getting rid of the fringing fin rays. But that alone isn't ideal: the appendage should be reasonably long to get decent leverage, and some kind of distal expansion would be handy to increase friction and prevent slippage. A vague ability to lightly grip would be useful, for similar reasons. The solution? Instead of simply doing away with the fin rays (leaving only the stubby lobe), why not additionally prolong the development of the fin's endoskeleton, tweaking the developmental pattern a little to broaden the far end?

This, in essence, is exactly what seems to have happened, although we're still a little hazy on the details. What we do know is that the development of the endoskeleton is coupled to that of the fin fold (the fin proper, containing the rays) in a mutually antagonistic fashion. Early on in the embryonic growth of the paired appendages, when they're little more than blobs, an encircling band of tissue forms around the end of each, called the apical ectodermal ridge (*ectoderm* being the surface cell layer of an embryo). The ridge is the source of chemical messenger molecules, including fibroblast (fiber-making) growth factors or FGFs, which turn on the endoskeleton-building program in the fin/limb bud. In fish, the ridge quietly churns out FGFs for a time while the endoskeleton builds, and then starts to grow itself, expanding into the fin fold. In so doing, it ultimately blocks the passage of the FGFs, bringing the proliferation of the endoskeleton to a halt. In evolutionary terms, this means that any developmental tweaking that extends the earlier skeleton-building phase automatically truncates the growth of the fin fold, thereby killing two birds with one stone. The converse is also true: activating fin fold growth prematurely gives a truncated endoskeleton and a larger fin blade. This is what the ray-finned fish did—it's how they lost their fin lobes.

The conversion of a fin into an *Acanthostega*-like limb might therefore have been a simpler evolutionary shift than you might think. Once our ancestors had taken to plodding about on the bottom it was perhaps only a matter of time before natural selection got to work on their fins and started to push them in a more robust, limblike direction. Which leads us inexorably to another question: why did our ancestors start sniffing around the bottom in the first place? Before addressing this point directly, it's worth noting that prototetrapods were neither the first nor the last vertebrates to express a certain benthic curiosity. Our friend the epaulette shark is a case in point. Skates, rays, and flatfish are probably the most obvious examples, and while flatfish don't walk on the bottom, several skates and rays have converted parts of their pelvic fins into rudimentary legs. However, the most remarkable underwater striders must surely be the dumpy-looking frogfish and the unrelated batfish. Both groups have transformed their fins into strikingly limblike appendages. Frogfish have taken the limb analogy so far that they can even grip with their expanded handlike fin tips, and are quite adept at climbing submerged rocks and seaweeds.

Looking at these underwater pedestrians can give us some valuable clues as to why our distant ancestors went down the walking road. Aside from those fish that, like the epaulette shark, feed on organisms that live within the substrate (making a cheap, substrate-based mode of locomotion an obviously attractive prospect), a common theme of stealth and ambush runs through the ecology of the benthic striders. The batfish and frogfish are particularly good at this sneaky way of life. Like anglerfish, they carry a juicy, twitching lure that dangles just above their mouth: a mouth that can expand in a split second to engulf any unwary opportunists that let hunger or curiosity get the better of them. Being able to maneuver by walking is a useful skill in these circumstances. Swimming into position would be a dead giveaway, the great currents of water stirred up making the would-be ambusher as obvious as if it had just shouted "boo!" at the top of its voice. Imparting momentum to the substrate rather than the water—tiptoeing up to a victim rather than sculling—is a strategy much more likely to end in a meal. And the wonderful thing is that there's no need to abandon the old aquatic skills, for the tail is still there should the fish feel like going for a swim. Indeed, every one of the walking fish described so far lifts off if it has to get anywhere (or away from anywhere) in a hurry: walking on the bottom is no match for swimming if speed is required.

Back in the Devonian shallows, it's not difficult to imagine our erstwhile open-water ancestors turning to a life of concealment and subterfuge. Ambush predation seems to have been the dominant mode of life of many sarcopts, judging by the common expression of a high-acceleration pikelike morphology, with much of the fin area concentrated at the back end (just have another look at *Eusthenopteron*). The Late Devonian must have been a paranoid time. With this in mind, it makes perfect sense that the odd group here and there would try out a bit of substrate-hugging, especially when one considers the fringe benefits: in shallow, vegetation-choked waterways a push on the substrate coupled with the ability to manhandle the reeds would have made navigation considerably easier, and if the water was really shallow, a little push-up from the pectorals would have made air-breathing a breeze (as we'll see shortly, these aquatic tetrapods-in-waiting were most definitely air-breathers).

## LIMBS ON LAND

Thus did limbs appear. (Probably.) And at some point following *Acanthostega*'s divergence, an enterprising young near-shore tetrapod, perhaps tempted by the tasty terrestrial arthropods, or maybe just tired of the endless run-ins with ambush predators, realized that limbs weren't just good for walking underwater. All the creature needed to do was push through the water's surface to open the door to a new world. But if it was going to make the most of this golden opportunity, some changes had to be made, for as we saw earlier,

*Acanthostega*'s appendages were better suited to underwater paddling than terrestrial striding. For one, they were saddled with big, heavy hands and feet that would be tiresome to lug around on land. That's an easy enough problem to solve—just reduce the number of digits, retaining sufficient fingers and toes to give a decent contact area (as we can verify, five digits on each limb appears to have been the best compromise). Then there's the lack of true elbow and knee joints, without which *Acanthostega* was permanently spread-eagled. To appreciate the difficulties of this kind of limb, imagine doing push-ups without bending your arms at the elbow (by all means, try it). It can be done, but the required effort is excessive because the ground reaction force, acting on the hands, is far from the line of action of the overall weight force, which points straight down from the center of mass. The wide separation of these opposing forces gives a large torque, which must be balanced by the pectoral muscles (among others) to lift the body off the ground. A terrestrial tetrapod would therefore do well to bring its *points d'appui* closer to its center of mass. Simply shortening the limbs would do the trick, but much of their leverage would then be lost. That's not necessarily a deal-breaker—snakes manage without. But there's another option: pop a joint into the limb so that it can flex toward the ground, achieving the usual press-up posture. This gives the best of both worlds: the effort required to support the body is greatly reduced without compromising stride length.

The effort-saving pulling-in of the limbs didn't stop with the early terrestrial tetrapods. The trend was continued much later when the ancestors of mammals and dinosaurs independently adopted more upright postures, with the legs drawn in to lie directly beneath the body. Indeed, one could view our diagnostically human valgus knees as a further continuation of the same tendency. These changes all brought about further improvements in the economy of support and locomotion, but at the cost of decreased stability, which would help to explain why some terrestrial tetrapods, most notably lizards, which are generally small animals that scamper about over uneven terrain, have hung on to the old sprawling posture.

We're not quite finished yet. An often overlooked complication of sprawling locomotion is that when using limb retraction to push the body forward while the supporting arm or leg is flexed, the forearm and lower leg must rotate about their long axes (try it and see). This is where the one-plus-two bone pattern of the sarcopt fin/limb comes into its own. With two long bones in the lower limb, long axis rotation can be brought about just by allowing the pair to cross each other like partners in a dance. It's a wonderfully elegant arrangement, for the rotation can proceed without compromising the stable hingelike motion required at the wrist, ankle, knee, and elbow.

We're now in a much better position to address our earlier question: why was it that despite the many mini-invasions of land carried out by various fish groups, only the tetrapods gained full terrestrial supremacy? It could be argued that we just got there first, and quickly radiated to fill and thereafter block access to all available beginner terrestrial vertebrate niches, but I have a hunch that there's more to it than that—that some aspect of our biology gave us an unfair advantage over our ray-finned cousins. After all, the long incumbency of fish in the sea didn't stop tetrapods muscling in on *their* turf (well, surf). That crucial advantage is likely to be found in our sarcopterygian ancestry: specifically, the special sarcopt-type fin skeleton. With that single limb bone articulating with each girdle and the two rotation-permitting long bones beyond, we had the ability to achieve what for actinopts must be nearly impossible: the perfect marriage of strength and flexibility. Actinopts generally have exquisitely flexible fins, but the flexibility entirely rests in the control they have over their slender fin rays and the containing flimsy fin fold—the endoskeleton has been reduced almost to nothing. When robustness is called for, as in the mudskipper, that lovely flexibility is lost, and the formerly balletic beauty of the fin is reduced to an industrial back-and-forth trudge. Not so for the sarcopterygian tetrapods. Thanks to our retention—indeed, expansion—of the endoskeletal component of each fin, there was no need for any compromise, as we can see to this day. We can (nominally) use the same appendages to both play and carry pianos.

What all this tells us is that as far as the tetrapod colonization of the land is concerned, the origin of the sarcopt fin was every bit as important as was the later conversion of the fins to limbs. Our exploration of the vertebrate land invasion won't be complete until we have looked into this much earlier transition. Ironically, it's likely that the initial impetus for the transformation was the selfsame impetus responsible for the origin of the actinopt fin type. It was the moment when vertebrates started breathing air.

## OF BREATH AND BONE

That last claim may sound a touch outrageous. We naturally associate air-breathing with life on land, yet here I am suggesting that the development of this ability caused two independent evolutionary remodelings of vertebrate fins, both of which occurred long before the first would-be terrestrial colonists left the water. Why would a water dweller breathe air? Well, why not? On land, of course, air is the only thing we *can* breathe, but the converse need not be true. If they live close enough to the surface, fish have the option of breathing either air or water, or both. And many of them do—very many, in fact, and not just the semiterrestrial fish we looked at earlier. Electric eels, gars, armored catfish, bowfins, arapaimas, mudminnows, featherbacks, gouramis, loaches, and the sarcopt lungfish, to name but a few, are all air-breathers, as is our old friend *Polypterus*: what Geoffroy identified as a double swim bladder is actually a pair of lungs. In all, there are over 370 species of air-breathing fish, and according to Jeffrey Graham of the Scripps Institution of Oceanography, San Diego, the habit likely evolved independently in vertebrates anything between thirty-eight and sixty-seven times. We can be sure that the transition happened multiple times because there are many different kinds of air-breathing organ: conventional lungs are only the most familiar option. Mudskippers, for instance, have converted the lining of the gill chamber into a basic lung; the electric eel, more bizarrely, has a lung in its mouth (small wonder it has to stun its prey before eating it); the armored catfish has one in its intestines (so, yes, it passes wind to exhale). None of these different air-breathing organs could have been derived from each other: they must have been invented from scratch.

That air-breathing has evolved so many times in vertebrates indicates that the transition is either particularly easy or beneficial, or both. The benefits are indeed great: oxygen is about thirty times more concentrated in air than in even well-mixed water, and diffuses ten thousand times faster. Remember also that air is eight hundred times less dense than water, so not only does an air-breather require a lower ventilation rate than a water-breather (assuming equivalent metabolic rates), but each breath requires vastly less effort from the ventilatory pumps.

Why, then, don't *all* fish breathe air? As always, of course, there's a flip side. First, there's the sheer inconvenience of having to swim to the surface repeatedly, and it may be dangerous to stay up there, with death at the sharp end of beak and talon waiting on feathered wings (although that wouldn't have been a problem for the first vertebrate air-breathers, as beaks, talons, feathers, and wings were all in the distant future). Second, all that stuff about the physical properties of air versus water is a closed book to fish; they're not going to go on a skyward voyage of discovery because they've heard of a legendary realm where oxygen is vastly easier to get hold of. They would have to stumble on the air/water interface accidentally.

All of that means that the best places to look for air-breathing fish are shallow tropical lowland freshwaters or estuaries, and indeed this is where the vast majority of today's air-breathing fish are to be found. Here, the water is typically warm, poorly stirred, and full of decaying vegetation, with the result that the oxygen content tends to be rather low, particularly at depth. That last point is important, for it means there's a bottom-to-surface oxygen concentration gradient that a fish will naturally follow if it's feeling short of breath, bringing it inexorably to the air above regardless of whether it ever intended to break the surface. As the water is shallow (shallower than an ocean, at least), the journey won't be a long one. Given the low oxygen content, a breath will help even if a fish has no anatomical modifications for air-breathing. Merely swallowing air will do the trick:* all parts of the gut, from mouth to anus, are well supplied with blood vessels, and can draw some oxygen from a gulp of air, albeit inefficiently. Obviously, natural selection can be expected in this situation to get to work on that inefficiency, dedicating a small part of the digestive tract to the new air-breathing function. And that's exactly what's happened, many times.

For whatever reason, sharks and their kin never turned to air-breathing, but their bony cousins seem to have adopted the habit very early on. Nearly all bony fish have an air sac of some description that branches off the gut between the mouth and the stomach (although the connecting tube may be lost during embryonic development). We tetrapods have our lungs, of course, as do our close sarcopt relatives the lungfish, while actinopts have their swim bladders. Granted, there are differences between the sarcopt and actinopt versions: sarcopt lungs are usually paired structures that grow from the lower side of the gut, whereas the typical actinopt equivalent is a single sac that sits above it. However, both kinds of air sac are lubricated with the same chemical solution and have a similar blood supply, raising the distinct possibility that the organs are homologous. Furthermore, *Polypterus*—the most primitive living actinopt—has a sarcopt-type lung: as you may remember, it's paired, and branches from the lower side of the gut. The weight of evidence therefore points toward a very early origin of sarcopt-type lungs in the bony fish lineage—before the two main branches went their separate ways—with a later conversion in the actinopt line. That means that the first breath of air

* Gills generally don't work well as air-breathing organs, because the flimsy filaments collapse and stick together when robbed of the support of water. *Synbranchus*—a kind of freshwater eel—is exceptional in this regard: its gill apparatus is strategically welded into a semirigid basket, so the filaments work as well in air as underwater.

may have been taken up to 490 million years ago—presumably because the ancestors of the bony fish found themselves in oxygen-poor water—and that all non-air-breathing bony fish have since lost the ability.

You might now be wondering what all this has to do with locomotion. The thing is, taking a gulp of air is not going to be without consequences, and I'm not just talking about the respiratory ones. Because air is so much more rarified than water, even a small volume inside the body—whose tissues tend to be slightly denser than water overall—has a big influence on an animal's buoyancy. Specifically, it becomes easy for a fish to match its overall density to its surroundings, such that it displaces its own weight in water and thus suppresses any inherent tendency to rise or fall. This property is known as neutral buoyancy, and to see the enormous influence that it or its lack of can have on fish, we need look no further than the group we started with. Lacking internal air sacs, sharks are as a rule negatively buoyant (they tend to sink), although only slightly, for they lack bone and typically have large livers packed with low-density oils. Indeed, both of these features are almost certainly density-reducing adaptations.†

Being negatively buoyant, sharks face the same problem that bedevils flying animals: to avoid sinking to the bottom, they must generate lift as well as the usual thrust that all fish produce. As has long been suspected but only recently confirmed in George Lauder's laser-illuminated flow tank, the chief source of lift is the tail.‡ This is typically asymmetrical, with a long upper blade that contains the vertebral column and a shorter lower blade that's supported only by fin rays. The skewed construction is the key to a shark's lift-making powers. As the tail is swung from side to side, the upper blade leads the lower so that if viewed from directly behind, it looks for all the world like a beating wing, and indeed it functions in the same way: because the tail fin meets the oncoming water at a positive angle of attack, the motion produces the all-important pressure difference between upper and lower surfaces. As you might imagine, producing lift right at the back end would run the risk of pitching a shark nose-down unless balanced by a bit of lift at the

† The implication is that sharks are secondarily, rather than primitively, boneless. While the endoskeleton was ossified following the evolutionary divergence of sharks and their kin, the first bone was, in essence, an exoskeleton, which covered the bodies of most early jawless fish.

‡ Skates and rays are different—when swimming, they flap their expanded pectoral fins to generate the required lift. Of course, being principally bottom dwellers, most of them aren't usually worried about sinking.

front. This is why sharks tend to be somewhat flattened—the entire body thus acts as a lifting hydrofoil.

Thanks to their early adoption of air-breathing, bony fish don't have to worry about any of this. For one, they have no pressing need for the density-reducing measures of sharks, so had no qualms about ossifying the originally cartilaginous endoskeleton,* as is immediately apparent from the group's name. When neutrally buoyant, continuous lift is unnecessary, so the tail gradually lost its asymmetry in both sarcopt and actinopt lineages. In the sarcopts, the back end of the vertebral column was straightened out and a new set of fin rays added to its dorsal side (nicely illustrated by *Eusthenopteron*), whereas the actinopt vertebral column retreated from the tail, turning the existing set of fin rays into a symmetrical fan.† Without the need to generate balancing lift at the front, the body was free to adopt all manner of shapes as various selection pressures demanded.

The acquisition of neutral buoyancy was so useful to the bony fish that it was retained even when the air-breathing function was lost—notably in marine fish for which repeated journeys to the surface would be impractical. It is in this context that the swim bladder lost its primitive connection to the gut. Those fish that inherited such a gas sac can no longer fill it by gulping air; instead, a specialized gas gland secretes oxygen into the swim bladder from the blood (ironically, what was a source of oxygen has thus become a sink). The loss of the primitive respiratory function of the swim bladder was a necessary prelude to the subsequent evolution of the many exotic kinds of air-breathing organ mentioned earlier. Presumably, the ancestors of the various fish that sport these weird and wonderful lungs, having downplayed the respiratory function of their swim bladders, were later faced with the same oxygen problems that had caused the gas sac to appear in the first place. Confusingly, it seems to have been easier to invent a new air-breathing organ from scratch than to resurrect the respiratory function of the original lung.

---

* Bony fish not only ossified the endoskeleton but also retained the ancient bone exoskeleton, although it was subsequently reduced in most lineages, tetrapods included. We still have remnants, though. Our skull, for instance, is mostly exoskeletal, but long ago sank inward to join the endoskeleton.

† This doesn't mean that the tail of bony fish can't generate lift—just that lifting is no longer its default mode. Lauder found that the bluegill sunfish can precisely modulate the stiffness of its symmetrical tail to make it work like a shark's asymmetrical tail, should it need a bit of lift at the back end.

## A TALE OF TWO FISHIES

Now that we've dealt with the origin of air-breathing, it's time to return to the question I alluded to earlier: how did this transition unlock the door to land? In short, the bonus neutral buoyancy was the key. For a start, by enabling the ossification of the vertebral column, the removal of density restrictions underwater indirectly made support on land a future possibility. But the way that neutral buoyancy influenced the evolution of the paired fins was what ultimately decided which vertebrates would get to inherit the terrestrial realm. To see how, we first need to meet another unusual shark.

Cookie-cutter sharks—found throughout the world's warmer seas—are a far cry from the endearing epaulettes that began the chapter. With thick, rubbery lips and a cartoonish set of sharp, triangular teeth in their lower jaw, they're not the most attractive of their kind. They're only about a foot long, but don't let the diminutive size fool you—they have a particularly vicious way of making a living. The name comes from their appalling habit of suckering on to passing larger creatures and carving perfectly round chunks of flesh from their bodies. Nothing big is immune from their attentions—great white sharks have been found with cookie-cutter scars on their flanks, and whale carcasses have been washed up covered with hundreds of the round bites. Most astonishing of all, cookie-cutters have been known to attack nuclear submarines, and gained considerable notoriety in the 1970s when they nibbled through the rubber sonar housing of several vessels, effectively blinding them.

Uniquely for open-water sharks, cookie-cutters aren't active hunters—they hang motionless in the water in squadrons, waiting for an unlucky fish or whale to pass their way. Most sharks would simply sink if they tried to do this, but a cookie-cutter's oily liver is so enormous that the fish is neutrally buoyant. In response to this most unsharklike property, the pectoral fins have been reworked. Sharks, as a rule, have large, broad-based fins with a restricted range of movement. They function a bit like the fixed wings of an aircraft, generating hydrodynamic lift to control the creature's trajectory in the water.[‡] The neutrally buoyant cookie-cutters have no need of such huge inflexible appendages, so their pectoral fins have been reduced to narrow-based paddles.

‡Unlike an aircraft's wings, shark fins don't generate lift continuously (once again, we have Lauder's team to thank for that observation), so a more appropriate analogy might be the ailerons of an aircraft or the bow planes of a submarine. Both only work if the craft is moving.

As far as we can tell, the original bony fish fin was likely similar to what we see in sharks and probably had the same basic hydrofoil function. When the group became neutrally buoyant about 490 million years ago, they were thus presented with the same fin remodeling opportunities that would later be made available to the cookie-cutters. In the actinopt line, the fin endoskeleton was reduced in its entirety while the fin fold expanded, giving the ultraflexible ray fins that are the hallmark of the group. The sarcopts did things differently. Rather than shrink every endoskeletal element, they instead eliminated some parts of the endoskeleton while leaving other parts more or less unchanged: whereas the primitive fish fin had several basal elements articulating with the supporting girdle, sarcopts kept only one—the humerus in the pectoral fin, the femur in the pelvic—along with whatever distal bones articulated with it. Getting a narrow-based, more maneuverable fin thus seems to have been the main focus for the sarcopts, as it was for the actinopts, but it was achieved without the effective loss of the whole endoskeleton. That was the great hidden gift of neutral buoyancy: the strong but flexible fin skeleton that would eventually take our ancestors out of the water.

This is, of course, pure conjecture, and it would be nice to see a sarcopterygian fin in action to lend credence to the idea that neutral buoyancy and fin maneuverability go together in the sarcopts, as in the actinopts. Unfortunately, the fins of our closest fishy relatives—the lungfish—are rather too specialized to be instructive. What we could really do with is a representative of an earlier radiation, but that would appear to require the resurrection of the ancient dead. Incredibly, in 1938 just such a miracle appeared to happen, rather fittingly at Christmas, when a bizarre fish was caught just off the Comoros Islands between Africa and Madagascar. It was blue and bulky, about 5 feet long, and had the most unusual tail. This was thick and fleshy, and the spine within (which was later found to be almost pure notochord) ran in a straight line right to the tip, where it carried a miniature fin like a tiny leaf. Tails like this had been seen before, but not in any living creature. Rather, they were known from a group of extinct fish—the coelacanths—which cropped up in the Devonian period and disappeared from the fossil record about 65 million years ago along with the (nonbird) dinosaurs. They were, in fact, one of Huxley's crossopterygian groups, and are now known to be one of the earliest-branching lineages within the sarcopterygian radiation. The realization that at least one species of this ancient dynasty had survived to the present day, when by rights they should have breathed their last at the end of the Cretaceous period, was the biggest zoological news story of the twentieth century, and given the pivotal position of coelacanths' divergence at the base of the sarcopt family tree, scientists the world over were desperate to see its fins in action.

The wait would be a long one. The living coelacanth is a reclusive, fairly deepwater animal that spends the day holed up in underwater caves, and quickly dies when brought to the surface. It would take the enterprising efforts of Hans Fricke, who from the late 1980s searched exhaustively for the fish in the purpose-built submersible *Geo*, to finally show us how the coelacanth behaved in its natural environment. Before these films were taken, it was thought that coelacanths used their lobe fins to walk around on the seafloor, but this isn't true: Fricke's footage revealed that they're committed swimmers, with a slow but beautifully graceful style that no shark could replicate. They can turn on the spot, swim backward, roll upside down, and do headstands with ease. In short, they're masters of low-speed maneuverability, and it's all thanks to the wide range of motion of their narrow-based lobe fins, which are continuously active during most of their movements. When looking at the footage, it can seem as if each fin has a mind of its own, but their strokes are in fact tightly coordinated to make sure that any unwanted accelerations brought on by the motion of one fin are canceled out by the concurrent motion of another. This usually means that the left pectoral fin and right pelvic fin retract together and vice versa. That might sound familiar—it's the same pattern that you'll find in a walking trot. Before we get too excited by the implications of this unexpected observation, we must remind ourselves that the Comoros coelacanth is a modern animal: we're not actually looking at one of our ancestors. Nevertheless, the use of an alternating gait in a fish with such an ancient pedigree means that we should at least entertain the possibility that the standard early tetrapod walking pattern dates right back to the very earliest sarcopterygians.

We finally have an answer to our central question: what was it about the tetrapods that gave them and no other vertebrate group the full terrestrial driving license? Simply put, we just had the right set of preadaptations. The ability to undulate the body was in place from ancient times. With the acquisition of air-breathing and neutral buoyancy came the opportunity to increase the flexibility of the fins and expand the locomotory repertoire. Whereas actinopts evolved themselves out of the full terrestrial running by effectively scrapping the fin endoskeleton, the sarcopt solution, which combined flexibility and strength, inadvertently kept the doorway to land open. With the possible early experimentation with alternating gaits for stability reasons and a penchant for ambush predation, it was only a matter of time before substrate contact occurred, followed by the expansion of the fin endoskeleton and the eventual appearance of limbs. A bit of vertebral consolidation here and a bit of limb joint refining there, and the tetrapods were ready to take on the rigors and opportunities of the terrestrial environment.

----------

Now that we're surrounded by them (as well as being among their number), it's all too easy to get the impression that the dry land was the vertebrates' birthright—that we were bound to make it ashore eventually. But such a notion seriously downplays what was in truth an utterly extraordinary transition. If you were to go back in time to the Silurian period, and got to know the vertebrates as they were then, the only reasonable conclusion you could draw would be that these were quintessentially aquatic creatures. As we saw in the last chapter, the very hallmarks of the group—the notochord and the vertebral column that ended up replacing it—are without question swimming adaptations. Even if the epaulette shark had been around at the time to plant the thought that undulations might be pretty good for walking, too, thanks to the presence of paired fins (also ostensibly swimming structures, of course), the very idea that such modestly built equipment would ever cut it in the heavyweight terrestrial world would seem ludicrous. And yet, in one of the greatest narrative twists of the history of life on Earth, we made it, all thanks to that first desperate gasp of a suffocating fish—an event reenacted with every breath we take—and the alchemy the swallowed air subsequently wrought on the vertebrate body.

6

# A Winning Formula

**In which we uncover the body-building blueprint that unleashed the locomotory carnival of the animals**

Truth is the shattered mirror strewn in myriad bits.

—Sir Richard Burton, *The Kasidah of Haji Abdu El-Yezdi*

For some species, the Darwinian thesis that all living things on Earth are related to us is a pretty easy sell. Chimps, gorillas, and orangutans are so similar to humans in so many ways—morphologically, physiologically, genetically, even psychologically—that one has to be in complete denial to reject the resemblance as sheer fluke. And despite their longer faces and typically quadrupedal postures, other mammals have the same warm blood, the same hairy skin (usually), and nourish their young with milk just as we do, so our mutual kinship isn't hard to appreciate. Were it not so easy to see something of ourselves in our mammalian cousins, it's doubtful that so many of us would keep them as pets. And while it might be more difficult to build a relationship with a mackerel or a shark, we plainly have much in common, from our beating heart to our trademark spinal column.

Looking further afield at our distant *in*vertebrate cousins, it can be more of a struggle to muster feelings of kindred affinity. From worms to spiders, snails to jellyfish, the body plans of the spineless hordes are so wildly different from ours—and indeed one another's—that we struggle to find any common points of reference at all. This is hardly surprising—the last common ancestor of insects and vertebrates may have lived 650 million years ago, and that of vertebrates and jellyfish up to 700 million. Even for the often glacially slow process of natural selection that's an awfully long time—enough to obliterate any overt traces of our shared ancestry.

Or so we might think. We're so easily dazzled by the morphological distinctness of the major groups of animals that we can miss what's staring us in the face: despite the vast reaches of time that separate us, our bodies tend to be organized along strikingly similar lines. For starters, the great majority of animal species have clearly defined front and back ends. Regardless of whether you're a vertebrate, an arthropod, a mollusk, or a worm, the mouth is almost always at the front, along with a collection of long-range sense organs and a brain (however modest), and the anus is usually at the back. Typically, left- and right-hand sides are near-mirror images of each other, whereas top (dorsal) and bottom (ventral) sides are usually quite different. Less overtly, almost all of us are at least partly built up of repeated, if potentially variable, units that are strung along the main axis of the body in a more or less regular array. Our vertebrae are obvious examples, but the same is true of our muscles (remember their primitive segmentation), nerves, teeth, and limbs. Among invertebrates, the modular design of arthropods, with their segmented armor and sets of paired appendages, is a particularly obvious example of this organizational logic, but even the lowly unsegmented flatworms betray an adherence to the same design principles: many of their internal organ systems—nervous, excretory, and reproductive structures, for instance—take the form of chains of repeated elements running down the body.

The left/right bilateral symmetry system is important enough to give its name to the group of animals that use it—the Bilateria. Together, we account for about 99 percent of all animal species, which clearly indicates that our basic blueprint is not just tenacious but adaptable in the extreme. It's also the sort of thing that's easily taken for granted. We're all so thoroughly accustomed to animals' left/right symmetry and distinct front and back ends that we hardly ever think to ask why the system is so common. Should we do so, however, we find that, on one level, the answer is quite straightforward. We have only to bear in mind an obvious fact—that animals are the only multicellular creatures that go in for directed movement in a big way—for the merits of the bilaterian design philosophy to quickly reveal themselves. As soon as an organism sets off, it cannot help but polarize itself—it instantly acquires a front end and a back end—and as long as the creature is consistent in its orientation, it will likely benefit if those two ends are specialized along different lines. The front, of course, will encounter the environment (including food) first, so it makes sense for the mouth and long-range sense organs to be located there. Conversely, waste is best discharged into the wake if an animal doesn't want to be continually moving through its own poop, so the best place for the anus is usually the back end. Symmetry about the fore-to-aft axis makes course-holding a much simpler task than it might otherwise be: if all the legs were on one side, for instance, an animal would find it very difficult to do anything but go around and around in circles. Finally, the repeated growth of propulsive modules along the main axis of the body—be they legs, paddles, or even just muscle segments (with their associated nerves)—will potentially increase the available thrust without disrupting balance.

So, there you have it: because locomotion is of such obvious importance to animals, it's no surprise that the main anatomical scheme seen in our kingdom is one that reliably yields efficient and effective locomotory bodies, regardless of the style of movement or the environment in which that movement takes place (everything said in the previous paragraph should apply equally to a swimmer, a crawler, or a burrower, for instance). However, it doesn't take much probing to discover that such a superficial analysis barely scratches the surface of this phenomenon. First, it's one thing to realize the importance of the bilaterian system, but quite another to understand how it is consistently implemented across the animal kingdom in the face of some truly enormous differences in morphology. A second complication is that while non-bilaterally-symmetrical animals are comparatively rare, they do exist. Most are radially symmetrical: they have a clearly defined main axis, but there's no equivalent of the bilaterian dorsal and ventral sides, and the duplicated structural modules are arrayed *around* the axis rather than along it—consider, for example, the five arms of a starfish. Some of these animals are the descendants of standard bilaterian ancestors that abandoned locomotion, so given what was said earlier, their lack of bilateral symmetry—a secondary loss—isn't hard to understand.* Others, however, went their separate evolutionary ways long before the bilaterians evolved. These are the jellyfish, corals, sea anemones, and the like—a group called the Cnidaria—and the

---

* Of course, starfish, and indeed most of the other echinoderms (which are secondarily radially symmetrical bilaterians), do move around, albeit slowly. I'll come back to this curious situation in Chapter 8.

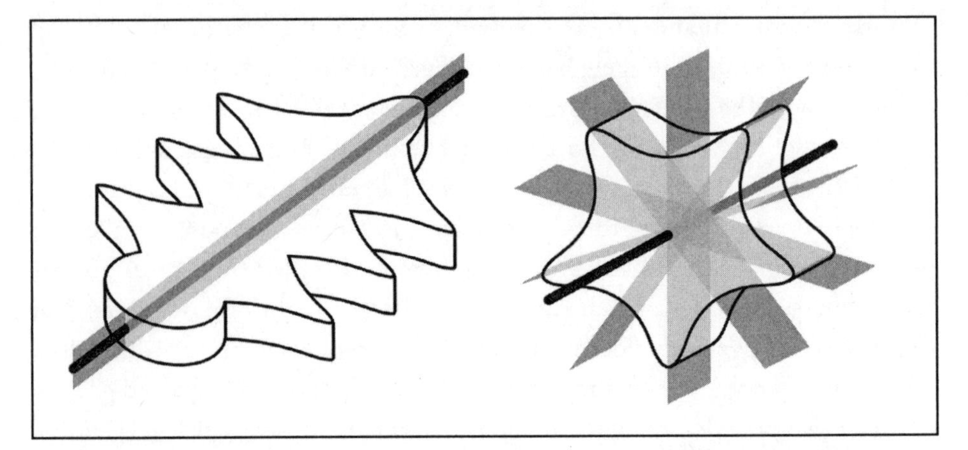

6-1: Bilateral (left) and radial (right) symmetry systems. A bilaterally symmetrical animal has a single symmetry plane running parallel to the main axis of the body, with duplicate elements repeated *along* the axis, whereas a radially symmetrical animal has several symmetry planes (five here) running perpendicular to the main axis, with duplicate elements repeated *about* the axis.

superficially similar but only distantly related comb jellies, or ctenophores. We could also include the even simpler sponges in the radial club—while these are often thought to be bereft of any kind of symmetry, quite a few are radially symmetrical, and even the truly asymmetrical representatives tend to be built up of repeated radially symmetrical bits. These various descendants of the prebilaterians present more of an enigma, for while many are rooted to the spot as are most of the secondarily radially symmetrical bilaterians, plenty are fully capable movers. And why shouldn't they be? I argued earlier that the high locomotory quality of a bilaterian body rests in its clearly defined front and back and its balanced array of propulsive modules. All that is true of a radially symmetrical creature, too.

This is starting to look rather mysterious. The prevalence of radial symmetry among the basal members of the animal kingdom tells us that this system is more primitive than its bilateral counterpart, and that the latter must have arisen from the former. But why, if both are perfectly effective templates for locomotory bodies? And why did the switch from radial to bilateral symmetry result in such an explosion of locomotory diversity and complexity? The jellyfish and comb jelly groups are pretty much one-trick ponies, while the bilaterians move around in all manner of different ways, from swimming to burrowing, running to flying. In this light, the conversion from one organizational system to the other is revealed as nothing less than the making of the animal kingdom; if it hadn't happened, jellyfish would now represent the pinnacle of locomotory achievement on Earth. Why? Finally (and this may be the most important question of all), how did the advanced bilateral blueprint evolve?

Answering these questions is going to require a different approach to the one we're used to. Up to now, when attempting to explain how and why certain locomotory changes occurred, we've focused our attention on the physical impact on movement of the appearance of certain anatomical traits, such as wings, opposable thumbs, the vertebral column, and limbs. Such traits are usually easy to define: excepting the existential fuzziness that always occurs at the evolutionary origin of new structures (how big does a wing have to be before you call it a wing, etc.), their basic natures are manifestly obvious. But if we are to work out why certain changes to an organizational system occurred, we need to stop and think, for it's not at all obvious what an organizational system amounts to in terms that are biologically meaningful. It's all very well speaking of blueprints and plans, but these are metaphors: animals aren't assembled on a workbench from a collection of components according to some technical drawing. We're living, breathing, and, most crucially, growing things, not flat-pack furniture. That's our clue: in reality, a biological blueprint is the way an organism grows from egg to adult. If we are to find out how and why the bilaterian system came into being, and why it has proved to be such a rich source of locomotory innovation, we must first come to grips with the strange and beautiful world of biological development.

## CHICKEN AND EGG SITUATIONS

The development of a complex organism from a single fertilized egg cell has to be one of the most wondrous of biological phenomena, and one of the most difficult to understand. Somehow, from the near-homogenous collection of cells that characterizes the early stages of embryonic growth, an exquisitely ordered structure unfolds without any input from an external craftsman. The whole process seems to border on the miraculous, as if the normal laws of cause and effect have been put on hold. Small wonder it's taken us most of our recorded history to work out the basics.

Aristotle, as usual, was one of the first people to think hard about development. His observations of freshly laid chicken eggs led him to conclude that as they contained no preformed chick, an animal's form must arise gradually, guided by its soul.* Fast-forward to the eighteenth century and such talk of vital spirits had become rather distasteful to the materialist thinkers of the Enlightenment. That meant, however, that they were faced with a chicken-and-egg problem in the figurative as well as literal sense. If an embryo couldn't guide its own development, what did? A popular idea, which went by the name of preformationism, was that development entailed only growth, and that a minute, prefabricated organism did exist, in the egg or sperm, prior to fertilization.

The discovery of the genetic basis of life in the twentieth century offered a middle ground between Aristotle's gradual assembly ideas and the preformationist theories of the Enlightenment. While no one at this point still believed that a mini-organism sat ready to go in either sperm or egg, it was hereafter clear that the information needed to build a creature did reside therein. But a major puzzle remained to be solved. When a cell divides, its entire complement of DNA is replicated, and each daughter cell inherits a copy (no gender is implied by "daughter," by the way—this is just what we call the results of a cell division). That means that every cell of a developing embryo contains exactly the same set of genetic instructions. How, then, can different kinds of cells be specified, or assembled into tissues and organs?

The simple answer is that no one cell in a multicellular organism uses its entire set of genetic instructions. In humans, for instance, only about 40 percent of our twenty-one thousand or so protein-coding genes are universally expressed. These are the so-called housekeeping genes, whose proteins are required for the routine maintenance and function of all cells (the various enzymes—protein catalysts—involved in obtaining energy from the breakdown of glucose fall into this category). The other genes are specific to a certain cell type or range of types (the protein insulin, for example, is made only by pancreatic cells). Something must act to turn such specific genes on or off in different cells. That something is a class of proteins called transcription factors, on account of their authority over whether a given gene is copied—transcribed—into a complementary strand of messenger RNA (mRNA); once complete, the mRNA strand leaves the nucleus and acts as the molecular template for the construction of the encoded protein from its amino acid building blocks. Transcription factors work by binding to a short stretch of

* Aristotle's concept of the soul was more complicated than the Christian version, with five progressively more refined flavors: all living things have at least a nutritive soul; animals add appetitive, sensitive, and locomotive properties; humans alone have a rational soul.

DNA associated with the target gene—the gene's switch—and either enable or veto its transcription by interacting with the various enzymes responsible for building the mRNA strand. These switches are often complex: one may contain binding sites for over twenty transcription factors, some acting as "on" signals, some as "off," so it might be helpful to think of them as combination-coded locks.

Regardless of such details, it's clear that we must turn to the transcription factors to explain how certain parts of the body become distinct from other parts—for example, how the cells of the front end develop differently from those of the back end. But transcription factors are proteins, which means they're encoded by genes, and if every cell has every gene, every cell can potentially make every transcription factor. Of course, one can invoke further transcription factors to turn *those* transcription factors on or off, but that leaves us in exactly the same chicken-and-egg cul-de-sac.

The beginnings of a solution to this conundrum were provided by German embryologist Hans Spemann and his PhD student Hilde Proescholt (later Mangold), who conducted a series of ingenious experiments on newt embryos in the early twentieth century. They found that a small region on the surface of the developing embryo had organizing properties such that if grafted onto another embryo, it initiated the formation of a second conjoined individual. This organizing center was identifiable even before the first cell division, as Spemann discovered when he used minuscule nooses fashioned from his baby's hairs to split fertilized newt eggs in two, artificially creating pairs of twins. If both half-embryos contained a share of the organizer, they gave rise to two perfectly normal-looking (if diminutive) tadpoles, but if the cleavage plane was so oriented that one twin got all the organizer and the other none, only the former became a tadpole—the other ended up as an undifferentiated, tumorlike blob.

What was it about this nondescript patch of a newt's fertilized egg that gave it such authority over the animal's future form? The full answer to this question would be a long time coming, but Spemann and Mangold uncovered a crucial piece of the puzzle when they observed that the organizer marked the site from which the embryonic gut would later form in a process known as gastrulation. This enormously important event, which occurs in all animals, with the possible exception of sponges, is the point at which an embryo starts to become morphologically interesting—something more than a simple ball of cells. It's at this stage that the growing animal organizes itself into germ layers: two in the cnidarians and ctenophores—endoderm and ectoderm—three in the Bilateria, which add mesoderm.* The endoderm forms the lining of the gut and its derivatives (including our lungs and liver), the ectoderm gives rise to the epidermis and neural tissue, and the mesoderm makes everything in between, including muscles and, in vertebrates, most of the skeleton, including the all-important notochord.

The specifics of gastrulation vary from group to group, but it always involves an inward migration of cells fated to become endoderm and mesoderm. In Spemann's newts (and many others) it's a process of invagination, whereby what has up to that point been a hollow ball of cells starts to turn in on itself, rather like what happens if you poke a slightly flaccid balloon. The resulting tube is the future gut, and its rim—where your finger is swallowed by the balloon, if we're sticking with the analogy—is called the blastopore. In newts (and all other chordates), the blastopore becomes the anus. As gastrulation proceeds, the embryo grows out from the blastopore end, turning the originally spherical organism into something more cylindrical in shape. In this way, the organizer (which, remember, marks where the anus will be) directs the foundation of the fore-to-aft axis of the newt from its posterior end, effectively telling the young newt what direction it's going to be moving in for the rest of its life. But there's more. The blastopore doesn't form right in the middle of the organizer, but just off to one side, so that the organizer ends up sitting on one edge of the future anus, not all the way around it. As the coupled processes of gastrulation and posterior growth proceed, the organizer is stretched into the interior of the embryo, forming a strip of tissue on one side of the gut. That side will be the dorsal side, and the strip of tissue will eventually become the newt's notochord. So, the organizer sets not only the fore-to-aft axis, but the dorsoventral axis, too: from a locomotory perspective, it both points the embryo in the right direction and makes sure its legs all end up on the same, ventral side—opposite the future vertebral column.

With the basic role of the organizer established, attention turned to the next big questions: how is a small part of the embryo singled out for the job of body-building project manager, and once promoted to this position, how does it carry out its organizing tasks? Spemann and Mangold's work uncovered a clue to the first of these questions: a newt's organizer doesn't form in a random location, but always lies directly opposite the point of entry of the fertilizing sperm cell. That immediately suggested that some message—likely chemical

* Sponges have distinct inner and outer layers of cells, but whether these are homologous to nonsponge ectoderm and endoderm is an open question.

in nature—is delivered by the sperm. This is indeed the case, though it would take many years before the nature of the chemical signal was identified. We now know that the signaling process initiated by a sperm's arrival is predictably complicated and indirect—involving the rearrangement of the egg cell's internal molecular scaffolding—but the upshot is the accumulation of a protein—a member of the Wnt (pronounced "wint") protein family—on the far side of the embryo. It is this protein that promotes the cells in the vicinity to the rank of organizer, whereupon they get on with their first task—kicking off gastrulation.

The Wnt protein can thus be thought of as a message, in chemical form, reading "this is the rear." Given its role in making the cells at the future back end distinct, you might think that it must be a transcription factor, but this isn't the case. Wnts are members of a different class of molecules, most of which are proteins, called morphogens, meaning "shape-makers." Unlike transcription factors, morphogens don't bind to DNA, but (usually) to special receptors—each specific to a particular morphogen as is a lock to a key—located on cell membranes. When a morphogen binds, a cascade of events is set into motion that results in the activation or repression of one or more transcription factors within the cell. While this might sound needlessly bureaucratic—why not just ditch the Wnt middleman?—morphogens are vitally important in development, for the simple reason that transcription factors almost always stay within their cell of origin, and so can't directly affect the fate of any other cells. Morphogens are more adventurous—most are released from their source cells and diffuse through the fluids of the embryo, binding to their target receptors as they go. The response of the target cells—that is, which genes get switched on or off therein—depends on which particular morphogens are bound, and also on the concentration of those morphogens, which obviously drops with increasing distance from the source. Simply put, morphogens are like short-range, developmental hormones, whose differing concentrations in different parts of the embryo bring about cellular differentiation through the control they exert over the cells' transcription factors.

In the case of the newt's organizer, some of the genes that are turned on by the high concentration of Wnt at the back end are responsible for initiating gastrulation. Others stimulate the embryo's posterior extension. Still others are genes for more morphogens, whose respective activations elicit subsequent phases of development, domino fashion. For instance, the extruding back end of the embryo churns out Wnt proteins as it goes, and so continues to mark the back end of the fore-to-aft axis. Other morphogens are produced by the organizer tissue as it grows into the embryo (the tissue that will become the notochord). One such morphogen is called Chordin (after the notochord): its ultimate role is to turn on various genes that are appropriate for the dorsal side of the newt, where Chordin is concentrated. The overlying ectoderm, for instance, is induced to become neural tissue. The morphogen's role is easily verified: inject Chordin into a different part of the embryo and a second spinal cord begins to form there. Its method of action is somewhat unusual. Rather than binding to receptors itself, it binds to certain other morphogen molecules (technically, it antagonizes them), and so prevents them from binding to their targets. The antagonized morphogens in question are members of the BMP—bone morphogenic protein—family that are expressed all over the embryonic mesoderm and ectoderm. They induce ventrally appropriate cell fates wherever Chordin (and other organizer morphogens) are absent.

If this sounds complicated, bear in mind that newts are complicated creatures. However, at the heart of the whole convoluted process lies an elegant chemical GPS, whereby cell identities are defined with respect to the major axes of the body—fore-to-aft and top-to-bottom—by two concentration gradients of morphogens, ultimately set up by the organizer, which lie at right angles to each other. The local Wnt concentration gives the *x*-coordinate, the local BMP concentration the *y*-coordinate. These coordinates are translated (via morphogen receptors and their signaling pathways) into location-specific transcription factors that unlock the combination coded switches of genes appropriate for the given address (Fig. 6-2). In this way, the right kinds of cells differentiate in the right place, and our newt grows up to be an effective self-propelled organism.

## THE UNIVERSAL INSTRUCTION MANUAL

At this point you may be wondering why we're looking so closely at newts—what can they possibly tell us about the ancient locomotory blueprints of our distant ancestors? As it turns out, the molecular instructions that specify the main body axes of amphibians, and thereby ensure that all the right bits are in the right places, are astonishingly widespread in the animal kingdom. A whole range of bilaterians, from flatworms to humans—regardless of their radically different morphologies—use Wnt gradients to determine their fore-to-aft axis, and BMP gradients for their top-to-bottom axis. The discovery that such distant relatives use the same fundamental patterning system came as quite a shock: imagine making first contact with an unknown human tribe, only to find that they're all using much the same vocabulary as you. The similarities of the axial specification systems are in many cases so close

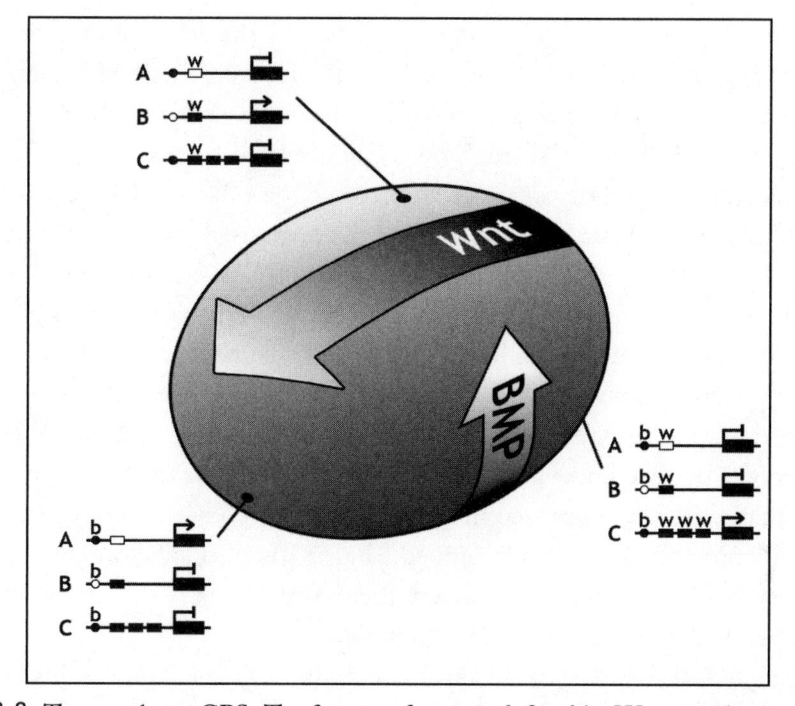


6-2: The vertebrate GPS. The fore-to-aft axis is defined by Wnt morphogens, whose concentration is high at the rear and low at the front. The dorsoventral axis is defined by BMP morphogens, whose concentration is low dorsally, high ventrally. The expression pattern of three hypothetical genes—A, B, and C—in three different locations of the body illustrates how these morphogen concentration gradients are "translated" for the purposes of cell differentiation. Each gene has a regulatory switch on the left and a coding sequence on the right, which is transcribed into messenger RNA when the switch is activated. The switches of all three genes have binding sites for the intracellular transcription factor "emissaries" of BMP (b) and Wnt (w): circles and rectangles, respectively. All black binding sites of a switch must be occupied by the respective transcription factor for the gene to be activated, but if a white site is occupied, the gene is switched off regardless. Gene A has an activator site for BMP but a repressor site for Wnt, and so is only switched on in the ventral anterior region of the body, where BMP is plentiful but Wnt absent. Gene B, on the other hand, has an activator site for Wnt but a repressor for BMP, and so is only switched on dorsally in the rear half of the animal. Gene C has an activator site for BMP, and so is ventrally expressed, and three activator sites for Wnt. With three sites, all of which need to be filled for activation, a very high concentration of Wnt is required for the gene to turn on, so C is only expressed right at the rear end of the animal.

that the relevant genes can be fully understood by the embryos of ludicrously distant relatives. For instance, if the mRNA of a fruit fly's BMP antagonist is injected into the belly of an embryonic frog, a second (frog) nerve cord is formed there, just as if the frog's own Chordin had been injected. This can only mean that the last common ancestor of the bilaterians—often known as the ur-bilaterian, or Urbi for short, must have used the same Wnt/BMP GPS. The fact that these genes have changed so little in separate lineages hundreds of millions of years after they diverged tells us that natural selection must have a phenomenally iron grip when it comes to body organization. While various genes acting downstream of the axis-specification system have changed beyond recognition—giving us the kaleidoscopic diversity of bilaterian bodies that we know and love—the vast majority of the countless mutations that must have appeared in the fundamental body-planning genes over the millennia have not been tolerated; that is, the resulting organisms have not survived long enough to replicate themselves. Such is the importance of effective locomotory design to the Bilateria.

Granted, there are some inconsistencies. In our own chordate lineage, for instance, BMPs induce ventral identities as we've seen, while their antagonists, like Chordin, specify dorsal tissues, such as the spinal cord. However, in the other bilaterians, everything is literally topsy-turvy: BMPs induce *dorsal* structures, and it's the *anti*-BMPs that make ventral things. Look again, however, and you'll find that what we think of as dorsal structures are ventrally situated in the nonchordates. Most obviously, their main nerve cords, if present, run along the belly, not the back. So, it seems that the system is really the same: BMP antagonists induce the formation of neural tissue. All that's changed is that long ago we chordates decided to live upside down for reasons unknown.* Remarkably, this is exactly what Étienne Geoffroy Saint-Hilaire proposed back in the early nineteenth century, and while his ideas were lambasted at the time, he has at last been vindicated by twenty-first-century biotechnology.

So much for the axes; what of the other common locomotion-critical bilaterian features—the repeated structures, such as legs and muscle blocks—that are strung along our bodies? Are such arrays underpinned by shared developmental processes, too? Before we come to that, we need to know a little more about organ development. For the very simplest animals, specifying

* There's actually a little more to it than this. The mouth tends to be on the ventral, substrate-directed side of *all* free-living bilaterians, chordate or not. That means that the chordate mouth must either be an evolutionary novelty (the old one having closed up) or that it migrated to the opposite side of the dorsoventral axis long ago.

each constituent cell type with unique sets of whole-body coordinates can be sufficient to build a body in which everything is in its rightful place. However, if cells are to be arranged in organs, this top-down, micromanagement approach starts to look at best cumbersome and at worst unworkable. If every cell in our eyes, for instance, was determined only by reference to our global GPS, it would take only minor disturbances in the bodywide morphogen gradients to disrupt the organ's proper development, leading to badly shaped lenses, poorly aligned retinas, muscles attached to the wrong part of the eyeball, and so forth. It's far too crude—rather like attempting to assemble a circuit board while wearing boxing gloves. Far better to devolve responsibility and let the organs manage their organization themselves.

Implementing such delegation within the rules of developmental genetics is straightforward. All that's needed is for the particular combination of transcription factors expressed in the place where we want the organ to grow to switch on the genes for the morphogens that are required to build the organ. Let's say, for example, that a transcription factor expressed in a small patch of the body (let's call it Sibling Rivalry) turned on the genes involved in initiating the primary body axis like *wnt*.† This would happen if those *wnt* genes had switches with Sibling Rivalry binding sites. The result? A second body axis would grow there, just as if organizer tissue had been transplanted. Of course, animals generally don't want clones growing all over themselves (unless they reproduce by budding), but the logic holds whatever the effects of the locally activated developmental programs (or modules, as they're known in the field). Initiating a more modest outgrowth could, for instance, give the animal a serviceable leg.

The modular logic of development comes into its own when building repeated structures. To take a familiar example, the activation in vertebrates of a member of the fibroblast growth factor (FGF) morphogen family at two locations on each side of the body—specified by different $x$-coordinates but similar $y$-coordinates—results in the outgrowth of our four limbs: an anterior pair and a posterior pair. The same *FGF* gene is activated in those two locations because it has two switches that contain binding sites for the particular combinations of transcription factors found only in those two places (thanks to the particular concentration of morphogens in those places). I'll leave aside for now the point that the front and back pair usually end up looking different. Theoretically, there's nothing to stop the number of switches on a

† By convention, gene names are given in italics; their proteins in roman, with an uppercase first letter. A *wnt* gene encodes a Wnt protein.

module-initiating gene from increasing indefinitely—allowing the module to be activated anywhere—as long as there's sufficient information in the axial morphogen gradients to uniquely specify every desired location. Indeed, this is roughly how the famous developmental workhorse *Drosophila*—the fruit fly—specifies the locations of its segment boundaries.

As the number of repeated modules increases, however, this unique geographical specification begins to look rather unwieldy. Some millipedes, for example, have several hundred pairs of legs: do they really need a separate switch for every pair? No, they don't. If a body continues to elongate from its rear end while the module-activation process is still going on, additional copies can be added by using just one switch. The only extra thing that's needed is a morphogen whose expression is cyclical, causing its concentration to regularly rise and fall like clockwork. The system works like this: imagine a person walking along a path at a constant speed. She's carrying a device that bleeps at regular intervals, and every time she hears a bleep, she drops a magic bean that shortly thereafter grows into a beanstalk. Before long, the path will have been filled with a regular array of beanstalks. Replace the bleeper with a clockwork morphogen, the person with a nonoscillating morphogen that keeps pace with the growing rear end of the embryo, and the beanstalks with the repeated modules (e.g., legs), and you have a reasonable picture of a system that goes in vertebrates by the name of the segmentation clock. We use the system to partition the embryonic precursors of our muscle blocks and vertebrae. These paired mesodermal blobs—called somites—are successively pinched off, in a nose-to-tail direction, from the mesoderm that lies on either side of the notochord.

The various molecular components of the vertebrate segmentation clock are now well known. Our nonoscillating morphogen that keeps pace with the growing rear end is actually a boundary between the expression domains of two morphogens: one—an FGF protein—is concentrated at the posterior end of the embryo, like Wnt, and the other—retinoic acid—has its source at the front. The clockwork morphogen is a membrane-bound protein in the Delta family, which binds to receptors—proteins in the Notch family—on neighboring cells. Its expression begins at the rear end and sweeps forward as a wave. The oscillation occurs because every time the Notch/Delta pathway is activated by the binding of morphogen and receptor, inhibitors of the pathway are switched on as well (in chicks and possibly other creatures, these include a protein that goes by the wonderful name of Lunatic Fringe). The signal therefore switches itself off after a short delay. The inhibitors then promptly and spontaneously decay, allowing the signal to turn on again, and the cycle continues. The genes that are responsible for pinching off the somites are activated whenever and wherever the oscillating (Notch/Delta) and non-oscillating (FGF/retinoic acid boundary) signals coincide, so in theory the gene responsible for the pinching off needs only one switch, with binding sites for the transcription factors made by the two signaling pathways.

For a long time, the segmentation clock was thought to be a vertebrate-only system, mainly because *Drosophila* partitions its segments all at once, not sequentially. However, the basic elements of the clock, including the Notch/Delta pacemaker, have recently been found at work in the segmentation of spiders, centipedes, and cockroaches, which tells us that what we see in *Drosophila* must be a later specialization. It also raises the distinct possibility that some kind of Notch/Delta segmentation clock was present in Urbi, just like the Wnt/BMP double-axis system, and so forms another part of the universal bilaterian instruction manual. The ability to duplicate structures along the primary axis has clearly been of great importance to the bilaterians from the beginning, and as we saw with axis specification, this is likely to be because it makes for a more effective locomotory body. But in this case, the benefits go far beyond the obvious, for where movement based on multiple propulsive units is concerned, the whole can be much greater than the sum of its parts. To find out why, we should take a closer look at the group that has exploited the locomotory potential of the bilaterian developmental photocopying system like no other: the arthropods.

## LEGGING IT

Arthropods, by any token, are a phenomenally successful group of animals. For a start, the group contains more known species than the rest of the animal kingdom put together, by a very long way. Granted, most of them are insects, but even if insects didn't exist, the remaining representatives—the arachnids and their allies, the crustaceans and the myriapods (centipedes, millipedes, etc.)—would still account for nearly half of the animal kingdom. They can be found in just about every ecosystem on the planet, often in great abundance, and usually play dominant ecological roles in any community in which they're found. Much of this success is down to the supreme adaptability of the arthropod body plan—in particular, the adaptability of the repeated, jointed appendages that give the group its name. With these, the arthropods have found ways to navigate effectively in just about any environment that an animal might encounter, just like the vertebrates. Small wonder, then, that no matter how the body has been molded and remolded in the countless arthropod lineages, the appendages are a near-ubiquitous feature: only a few strange parasitic forms have got rid of them entirely.

Given the great diversity of arthropod bodies, we might expect to find a similar diversity in the way they use their appendages, but as far as locomotion is concerned, the basic method of operation shows surprisingly little variation across the group. Centipedes and millipedes, with their near-uniform sets of walking legs, illustrate the technique particularly clearly. As in all walkers, the legs cycle through stance and swing phases, during which they're alternately retracted while in contact with the ground to give the body a forward push (the power stroke) and protracted to reset. For obvious stability reasons, the legs don't act in perfect synchrony: if they did, the body would collapse to the ground every time they executed their swing phase. Instead, the step cycle of a given leg is usually slightly ahead of the one in front, such that the entire set collectively enacts repeated Mexican waves of activity—or metachronal waves, to give them their technical name—that pass from the rear end to the head. The coordination must be precise to avoid clashes between adjacent legs, but as long as this condition is met, the creature enjoys a remarkably smooth and economical ride, the bulk of the body maintaining a constant speed and height, with only the legs doing the fuel-guzzling accelerations and decelerations.

The repeated swimming legs of aquatic arthropods, such as prawns, are operated in much the same way as walking legs, with the same tail-to-head metachronal wave of activity, for similar reasons: moving the legs synchronously would give a jerky, stop/start character to the movement, whereas a metachronal pattern gives a steadier stream of water in the wake. As a result, a swimming prawn, like a walking centipede, is a picture of locomotory grace. In fact, if the spacing between the legs and the timing of the Mexican waves are just right, the locomotory benefits can be particularly fine: the periodic separation and approach of adjacent legs will alternately suck and squeeze water into and out of the spaces between, and because the water packets are drawn from the sides but expelled to the rear, this trick increases the wake momentum and can dramatically improve propulsive efficiency.*

* Given the near-ubiquitous use of metachronal waves in swimming crustaceans, have human rowing teams got it all wrong with their synchronous oar-stroke technique? Having spoken to some rowers on the subject, I've discovered that synchronizing the strokes delivers a hidden benefit—the hull rises a little during the power stroke, causing its drag to drop. Of course, that only applies if you row on the surface—a rare mode of locomotion in the animal kingdom.

It goes without saying that the ultraefficient metachronal pattern is only possible for an organism with regularly arrayed, repeated propulsive units; we've uncovered one almighty benefit of the bilaterian developmental photocopying system. But there's more. While the methods of operation of swimming and walking arthropod legs are much the same, the morphological design requirements are quite different. Walking legs are struts that must both support an arthropod's weight and push back on the ground, whereas swimming legs need to be flattened paddles that can be feathered during the reset (paddling is drag-based). For dedicated walkers (e.g., millipedes) or dedicated swimmers (e.g., brine shrimps), this is of little relevance, as their legs can all be of the same type. But many aquatic arthropods, such as prawns and lobsters, like to keep their options open, and so they have a set of each. Furthermore, many crustaceans keep a pair of broad, fanlike appendages at the back end: these so-called uropods are used in emergencies, when the tail is rapidly flicked back and forth a few times to accelerate the animal away from danger. The breadth of the uropods increases the momentum imparted to the water, just like a fish's tail. The options don't end there. A walking leg can be converted into a pincer by the growth of a long process on its penultimate segment: as well as the obvious cutting and crushing actions, such appendages can be used for grasping and climbing, just like our opposable thumbs. And regardless of whether they're terrestrial or aquatic, almost all arthropods employ varying numbers of modified legs as mouthparts (crabs have twelve!). This great diversity of appendages presents us with a puzzle. It's easy to see why it would be useful, but how do arthropods grow legs of such different designs, when all have been made using the same developmental module? Surely any mutation that influences the shape of one leg is necessarily repeated in all of them.

## THE SPICE OF LIFE

Up to now we've assumed that once activated, the repeated appendage developmental modules pay no attention to the GPS coordinates given by the Wnt/BMP gradients. But just because the modules *can* run without such input doesn't mean they have to. There's nothing to stop the local morphogen concentrations having a location-specific effect on a module's developmental trajectory. This is where a very special family of transcription factor genes called Hox genes—found in all animals except sponges—comes into play. Simply put, their job is to translate the fore-to-aft morphogen gradients into a transcription factor expression pattern. There are commonly between six and twelve of them in any one species, and in many animals they nestle together on a single chromosome in a tight cluster.

The Hox gene activation pattern, determined by the specifics of the genes' switches, serves to divide the fore-to-aft axis into a number of discrete bands, each defined by a unique Hox code—this being the collection of Hox transcription factors expressed in the band. This is the key to the variety of an arthropod's appendages, for while any repeated structures within one of the Hox regions will tend to look the same, the structures can take on an entirely different character, or may even be shut down completely, wherever the Hox code changes. For example, most centipede segments express the Hox genes *Ultrabithorax* (*Ubx*) and *Abdominal-A* (*Abd-A*), so the legs in this region all look the same. The segment nearest the head, however, expresses a different Hox code—*Sex combs reduced* (*Scr*) + *fushi tarazu* (*ftz*) + *Antennapedia* (*Antp*). Its pair of legs is accordingly quite different: they face forward and contain poison glands. These venom-laced fangs are the centipede's secret weapon, and provide a compelling incentive to shake out your boots every morning before hiking in the tropics, where centipedes get big (look up *Scolopendra* if you have a moment).

The presence of the Hox cluster alongside a system that builds duplicate appendages has truly profound evolutionary implications. Back in Chapter 3, when we first met the insects, I called attention to the evolutionary potential provided by having two pairs of wings. Because flight can work perfectly well with one pair, the other pair can be selectively tinkered with without messing things up—hence the appearance of halteres and beetle wing cases. What's true of insect wings is true of all arthropod repeated appendages: the duplicate sets give a measure of redundancy to the locomotory system (many arthropods continue to move around despite losing the odd leg), and as long as the developmental program of one set can be tweaked without automatically tweaking all the others as well, the stage is set for some serious evolutionary voyages of discovery. The Hox cluster makes this all possible: an arthropod can make as many different kinds of appendages as it has Hox codes. Furthermore, it is relatively easy for a region of the body to acquire a new Hox code: a shift in the boundaries of a Hox gene's expression domain will do the trick, which requires only that a Hox gene's sensitivity to the fore-to-aft morphogen gradients be adjusted by tinkering with its switches. Like a rock band member who signs with a new agent, an appendage that acquires a new Hox code can set off on an evolutionary solo career. With its

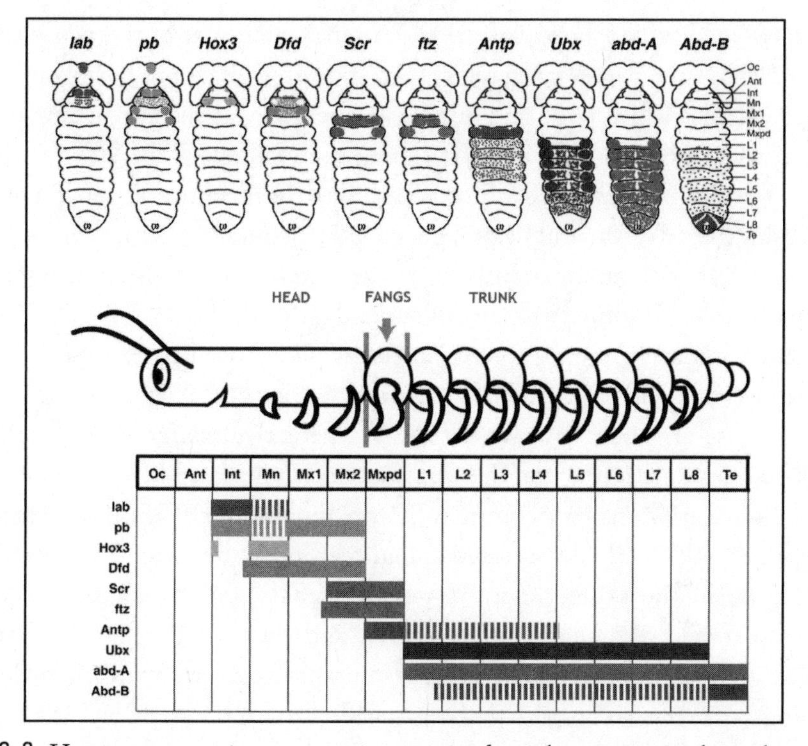


6-3: Hox gene expression patterns in a centipede as they appear in the embryo (top) and shown schematically (bottom). Thanks to the various fore and aft boundaries of expression in the Hox genes, different segments have different Hox codes, enabling the specification of diverse appendage types. The Mxpd (maxilliped) segment, for instance, uniquely expresses the Hox genes *Scr+ftz+Antp*, which is the code for poison fangs.

developmental fate now relatively isolated from that of the other legs, it's free to explore new morphological (and therefore functional) options, as the wondrous variety of arthropod appendages demonstrates to excess.

A while back I brushed under the carpet the common observation that vertebrate fore- and hindlimbs (or pectoral and pelvic fins) tend to be quite distinct despite being made by the same developmental module. This is no small matter: our freedom to push our fore- and hindlimbs in different evolutionary directions has been a major contributor to our success: bird wings, ape orthogrady, and the dramatically different adaptations of human hands and feet, for instance, could not have arisen otherwise. By now, however, there should be nothing troubling about the apparent contradiction. If Hox genes have allowed arthropods to turn their bodies into Swiss Army knives, why shouldn't vertebrates have followed suit? The only box that needs checking is the possession of a Hox cluster, and a quick glance at our genome shows that we've got the necessary genetic gear. Indeed, we seem to have gone overboard: whereas arthropods and our close lancelet relatives have a single Hox cluster, we've got four of them. At some point following the lancelet divergence, the protovertebrate genome was copied in its entirety, and then copied again. Some of the duplicated genes have degraded over time, but we've still been left with an apparently excessive number (thirty-nine in humans), although we've got only two pairs of appendages!

Thirty-nine Hox genes for two codes would indeed be excessive, but of course, we have other repeated structures besides our arms and legs. Chief among these are the derivatives of the segmented mesodermal somites: the axial muscles and vertebrae. The vertebrae in particular, like arthropod legs, are obvious variations on a common theme. Furthermore, again as in arthropods, there are clearly demarcated spinal regions within which the vertebrae tend to be of the same general type. Mammals, for instance, have well-defined cervical, thoracic, lumbar, sacral, and caudal (tail) regions. Studies of Hox gene expression in normal and mutant mice have confirmed that in keeping with the arthropod theme, the identity and extent of each region are down to the particular combination of Hox genes expressed there. The sacral region, for instance, is specified by the combined expression of *Hox10* and *Hox11* genes. Deleting the latter set gives the sacral region the same Hox code as the lumbar region, so a sacrum doesn't form, and the unfortunate mouse ends up with extra lumbar vertebrae instead. It isn't hard to find gentler examples of Hox-mediated vertebral tweaking. The fact that apes and humans have different numbers of lumbar vertebrae, for example (a discrepancy with important implications for spine stiffness and thus the origin of terrestrial bipedalism), is attributable to changes in the fore-to-aft extent of the expression domains of some of our Hox genes.

All in all, the way we vertebrates pattern our fore-to-aft axis is remarkably similar to the way arthropods do it. This isn't simply a case of using genes of the same family in a broadly analogous way. The vertebrate *Hox5* genes and their arthropod homologue *Sex combs reduced*, for instance, are all involved in patterning the back of the head and whatever lies immediately behind it. Despite our long separation, we're still using versions of the very same genes to build body parts in equivalent positions along our respective fore-to-aft axes. We clearly need to add the Hox-gene-based axial patterning system to our list of developmental processes that form the at-least-650-million-year-old universal instruction manual, alongside the Wnt/BMP GPS and some kind of Notch/Delta segmentation clock. For a group as outrageously diverse as the Bilateria, that's an impressive amount of common developmental ground. The time has come to put this knowledge to work. It's all very well pointing out how the bilaterian instruction manual has done wonders for the locomotory capabilities of vertebrates and arthropods, but if we're to understand why this scheme of development was selected in the first place, we're going to have to dig much deeper. It's time to find out about the life and locomotion of Urbi, our venerable ancestor.

## URBI GOES BANANAS

The developmental information gathered from arthropods and vertebrates tells us that Urbi almost certainly used a Wnt/BMP double gradient, and so had distinct front and back ends, along with top and bottom sides. It also had a Hox cluster—probably with six or seven Hox genes—and must therefore have had somewhat distinct body regions along its fore-to-aft axis. Given its likely use of a Notch/Delta-based segmentation clock, we might also suppose that it had duplicate units running down its body. But what exactly was it duplicating? Vertebrates and arthropods don't seem to be much help here, for their obvious repeated structures are fundamentally distinct: on the one hand, we have ectodermally derived, exoskeletal outgrowths (arthropod legs); on the other, a chain of mesodermally derived endoskeletal bones (vertebrae). Vertebrae are obviously vertebrate-only, and legs, while fantastically useful for the arthropods, are rare in the remaining bilaterians, being present only in the arthropods' close nonarmored relatives (including the superficially centipede-like velvet worms), as well as the segmented worm group—the annelids*—and the tetrapods, whose legs (and even the fins that preceded them) appeared long after the origin of the group and so cannot be homologous with what we see in arthropods anyway. It's vanishingly unlikely that Urbi had any of this stuff.

Before we get too disheartened, however, we should note that vertebrae and legs are not the only repeated things in vertebrates and arthropods, for both groups also partition their musculature. The vertebrates—and invertebrate chordates—show us that you don't need to have legs for segmented muscles to be useful. The undulatory swimming technique that was so critical

* Legs are commonplace in the marine annelids—their absence in earthworms reflects a secondary loss in connection with their terrestrial burrowing habit.

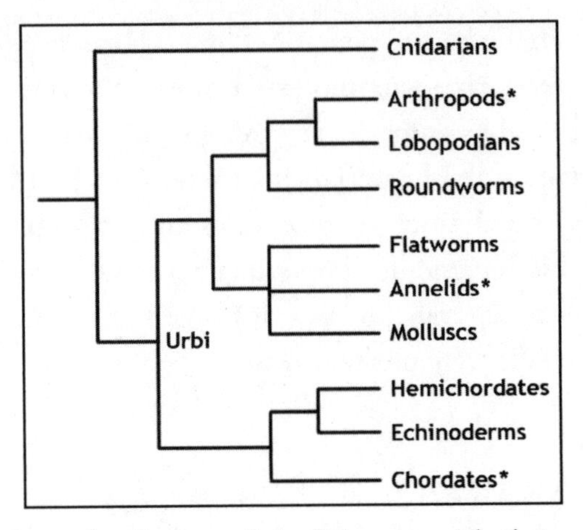


6-4: A pared-down family tree of the Bilateria, with their cnidarian sister group. Fully segmented groups are indicated by an asterisk.

to our success is based on the very same Mexican waves that are used to such great effect in arthropods.* And we are not alone. The peristalsis technique used by worms that we looked at in Chapter 4 is another manifestation of the same metachronal pattern, and requires a similar fore-to-aft partitioning of the longitudinal and circular musculature. Such partitioning doesn't necessarily result in overt segmentation. The muscular foot of gastropod mollusks, for instance—snails, slugs, winkles, limpets, and so on—appears to be made up of a single, homogeneous block of tissue. But when gastropods are on the move (and you can verify this by getting them to move on a pane of glass) waves of muscle contractions pulse across the foot. What they're doing is the closest an animal can get to walking without actually doing it. The sections of the foot in contact with the ground are retracting to provide the ground reaction force, while the sections between are protracting to reset.† Again, it's all done metachronally. The locomotory efficiency provided by those Mexican waves is clearly not to be underestimated.

* That's not to say that we vertebrates don't operate our limbs metachronally, too—we usually do, although with only two pairs of limbs, the Mexican waves are much less obvious. Nevertheless, trots, gallops, and bounds (even the alternating arm and leg swings of humans) are all metachronal patterns.

† In some species, the foot never actually loses contact with the substrate, which should make net movement impossible: resetting a foot section while it's still in contact should push the gastropod back where it came from. That this doesn't happen is down to the layer of mucus between the foot and the ground, which has an odd property known as thixotropy. It's normally a sticky solid, but if sheared hard it becomes liquid, thereby allowing the protracting sections of the foot to slide with respect to the substrate rather than pushing on it.

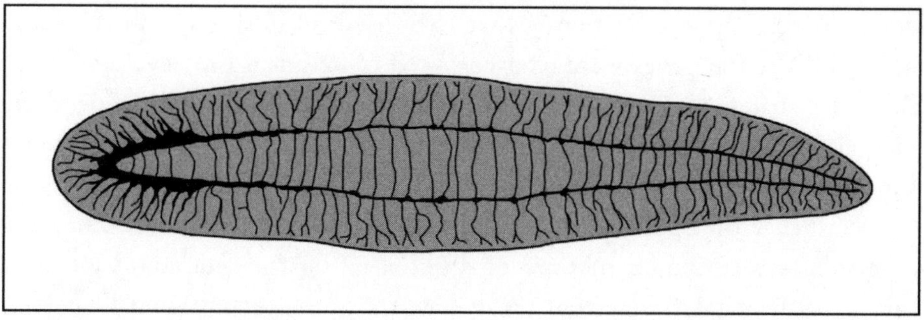

6-5: The segmented nervous system of a nominally unsegmented flatworm.

Given the common expression of metachronal wave–based locomotion in the Bilateria, we would be justified in assuming that Urbi made use of the same technique, and that it also had some kind of fore-to-aft partitioning of its musculature, although overt segmentation is unlikely. Flatworms are likely the closest modern analogues. These typically use beating cilia to glide on secreted mucus tracks, but they can also use muscular propulsion if, for instance, the viscosity of the surrounding fluid is increased. When they do, they use the familiar metachronal pattern to bring about either peristalsis or gastropod-like foot waves. Muscle bands might then be regarded as good candidates for the first repeated structures of bilaterians and the first target of the segmentation clock. But there's a second option. The best organized muscular system in the world won't do you much good if your motor nerves are all over the place, switching muscles on and off willy-nilly. What we could do with is an orderly network with, say, one or more bundles of nerves running from head to tail in longitudinal cables, and a series of left/right symmetrical lateral nerves branching off the main tracts at regular intervals to hook up to the musculature. This is exactly what we see and, crucially, we don't see it just in the chordates, annelids, and arthropods. Similar networks are seen in flatworms, some mollusks, roundworms—in short, the majority of the animals that use metachronal waves, regardless of whether their body shows any trace of segmentation, and regardless of whether they're walkers, burrowers, gastropod-like pseudo-walkers, or undulatory swimmers.

The evidence thus seems to indicate that the primary role of the segmentation clock was the axial organization of the neuromuscular system, and that this was in full swing in Urbi. Further support comes from the observation that despite the diversity of Hox-code-responsive structures in different groups (legs in arthropods, vertebrae in vertebrates), the use of Hox genes to pattern the nervous system along the fore-to-aft axis is the common denominator. The implication is that Urbi used its Hox genes in the same way, and that the genes were later co-opted to help run the development of the other structures in their various lineages, no doubt because in developmental terms, said structures were having to dance to the nervous system's tune anyway.

Whatever the specifics, one thing is clear. The combination of an axial repetition system and a method of differentiating the redundant locomotory modules that the system created was a potent evolutionary catalyst. The beginning of the Cambrian period—some 540 million years ago—saw an explosive diversification of the Bilateria, with almost all modern groups and their cornucopia of body plans appearing within about 25 million years. Much of this diversification was either directly or indirectly related to the locomotory options provided by the bilaterian blueprint. Surface gliders, which continued Urbi's fine tradition, were joined by burrowers, crawlers, scuttlers, clamberers, and swimmers (of drag-based and lift-based varieties), as well as jacks-of-all-trades that were able to combine multiple modes. As the locomotory facility of the various bilaterians improved, so, too, did their guiding sensory systems. Inevitably, some individuals began to turn their hungry eyes onto one another, and various forms of armor duly appeared, from the monolithic shields of most mollusks to the articulated plate mail of arthropods. That addition of an exoskeleton to the preexisting legged body plan seen in the arthropods' soft-bodied forerunners was a particularly significant development—the evolutionary equivalent of pouring water onto an oil fire—no doubt because of the expanded potential for complex, intricate leverage provided by the many pairs of rigid, jointed appendages. Small wonder that arthropod fossils are so thick on the ground in Cambrian deposits.

Despite all that, the bilaterians took their sweet time to exploit the locomotory possibilities of their developmental system. Urbi was around about 650 million years ago, going by the genetic distinctness of its modern descendants. That means it lived nearly 100 million years before the start of the Cambrian explosion. Why did the early bilaterians wait so long to make the most of the adaptability of their developmental toolkit? The simple answer: we're not sure. Environmental factors have been invoked—the lead-up to the Cambrian saw higher oxygen concentrations, increased input of calcium to the oceans (useful for skeleton-building), and rising sea levels (giving lots of new marine habitats), all of which may have removed the evolutionary brakes on bilaterian complexity. Or maybe the Urbi-grade developmental system wasn't quite sophisticated enough to generate the diversity of bilaterian bodies seen later in the Cambrian. A few gene duplications might have been required to give enough Hox codes to really get to work on diversifying the axially repeated neuromuscular system. Ultimately, it was probably a combination of factors that lit the fuse of the Cambrian bomb. Nevertheless, it cannot be denied that the core bilaterian developmental system and its locomotory potential were absolutely critical to how the Cambrian diversification played out: the bilaterian blueprint was the explosive, if not the spark.

## BROKEN MIRRORS

For all its violent connotations, the Cambrian explosion and its long adaptive aftermath barely touched the core bilaterian developmental program itself. Since it was invented in our distant forebears, it has only rarely been substantially altered, for it is still a supremely effective system for constructing fantastically efficient and adaptable locomotory bodies. It's no exaggeration to say that its locomotory potential allowed the bilaterians to inherit the Earth. We now arrive at our million-dollar questions: how did the world-shaking bilateral development system arise from the previous radial system, and why did the older system not unleash a similar locomotory explosion? To answer these questions, we're going to have to repeat what we did to piece together Urbi's developmental toolkit: look at our closest living nonbilaterian relatives to try to find some common developmental ground, and go from there. That means getting to know the Cnidaria. Now, some might regard these animals rather like an embarrassing uncle that no one likes to talk about: they're headless, brainless, have a mouth that doubles as an anus, and generally seem to bear no morphological resemblance to us bilaterians at all. But by now, we shouldn't be letting a little thing like complete morphological disparity stand in our way—not when we have developmental genetics at our disposal.

First, a little more about our gelatinous cousins. We recognize three groups: the anemones and corals (the Anthozoa) that are the more primitive of the three, the jellyfish (Scyphozoa), and the sometimes solitary, sometimes colonial hydroids (Hydrozoa). The basic cnidarian life cycle, seen in the anthozoans, has two stages: an adult polyp phase (an anemone being a typical if somewhat large example) and a mobile larval phase. The larva—called a planula—is a tiny egg-shaped creature that propels itself with cilia: beating, threadlike structures that cover its minuscule body. It usually swims around a bit before settling down on a suitable surface, whereupon it transforms into a polyp. In the advanced hydrozoans and scyphozoans, a third stage—the

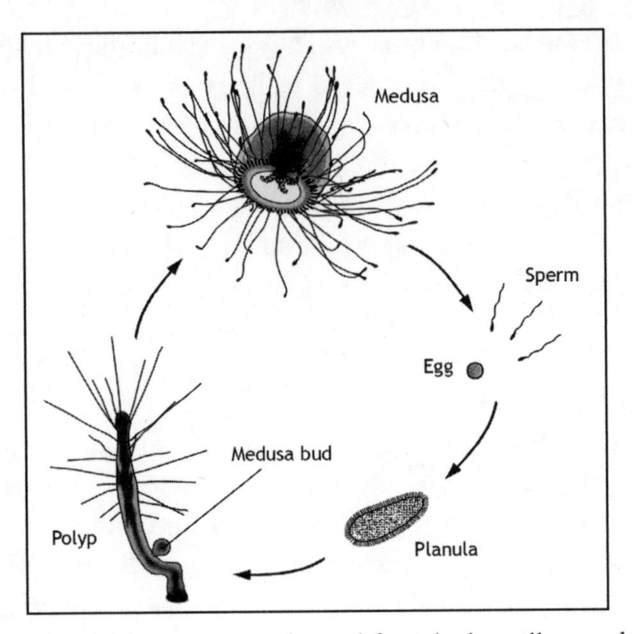


**6-6**: The advanced three-phase cnidarian life cycle, here illustrated with the hydroid *Turritopsis nutricula*.

medusa—is added (although many hydrozoans have since lost this phase). Medusae are swimming jellyfish-like animals (indeed, jellyfish in the strict sense are really just overgrown medusae) that usually bud off from polyps and carry out sexual reproduction. Being an advanced feature, medusae were not present in the last common ancestor of the Cnidaria and Bilateria, so are not directly relevant to our inquiry.

Now that the introductions are over, I have bad news: as far as Hox genes are concerned, the cnidarians (at least the few whose development has been looked at) are all over the place. Although putative Hox genes are present in these creatures, their evolutionary relationship to bilaterian Hox genes is highly uncertain, they're not arranged in a cluster, and while they show regional expression patterns, these vary greatly between species—even between life stages of the same species. In some cases, the expression patterns are distributed along the primary body axis (the axis that the mouth is at one end of), but not always. We're not going to be able to glean much useful information from this line of evidence, at least not for the time being.

Now for the good news: cnidarians have Wnt proteins, and here we seem to be on more familiar ground, for most are expressed at one end of the primary axis, just as in the bilaterians. That end turns out to be the mouth end, which means that, from a bilaterian perspective, polyps are standing on their head. While this might seem strange, it means that the bilaterian rear end equates to the rear end of a swimming planula. It is of course a planula's front end that makes first contact with the substrate, so it makes sense that the future polyp mouth opens at the other end. Just as in bilaterians, Wnt proteins therefore tell a cnidarian's body what direction it's going to be moving in during its short period of locomotory freedom. The role of this morphogen family in making sure the body is properly organized and oriented for movement appears to be truly ancient. Indeed, this function may date back to the very beginning of the animal kingdom. Maja Adamska and her colleagues at the University of Queensland, Australia, recently discovered that lowly sponge larvae also express a Wnt protein at their back end, defined with respect to their direction of motion. No matter how different we bilaterians, cnidarians, and sponges may appear to be, we at least seem to sort out our primary axes in the same way. That said, in the anemone-like hydroid *Hydra*, various genes are expressed at the mouth end (i.e., the back end) that in bilaterians are normally expressed in the head, contradicting much of what I've just said. Even more confusingly, other head-specific bilaterian genes are expressed farther down *Hydra*'s trunk. What's going on?

Hans Meinhardt, a *Hydra* expert at the Max Planck Institute for Developmental Biology in southern Germany, has an intriguing theory that could account for these apparently conflicting lines of evidence. He thinks that the bilaterian head corresponds to the entire body of cnidarians, excepting a narrow ring around the mouth, and that the origin of bilaterians from a cnidarian-grade animal was therefore associated with an elongation of that narrow band. To put it another way, bilaterians evolved their wormlike body by pulling their head out of their anus. As we know, an extrusion of the trunk from its posterior end—driven by Wnts—is a standard part of the core bilaterian body-building program, which lends considerable support to this suggestion, as does the lack of correspondence in bilaterian and cnidarian Hox gene expression patterns: we chiefly use them to pattern the trunk.

If the primary body axes of cnidarians and bilaterians are homologous, as Meinhardt's work suggests, our next task is to find out whether cnidarians have any hint of an axis at right angles that might have given rise to our dorsoventral patterning. As it happens, quite a few of them do. Many anthozoans, in particular, are not as radially symmetrical as one might suppose, with long, slitlike mouths. An additional body axis can therefore be described, referred to as the directive axis, to avoid any overhasty bilaterian homologizing. Suggestively, homologues of the bilaterian dorsoventral axis specifiers BMP and Chordin are expressed asymmetrically along the directive axis of sea anemone larvae. However, they're all expressed on the same side of the body, whereas in the Bilateria, the BMPs are on one side, Chordin on the other. Even more confusingly, the expression domains of some of the putative cnidarian Hox genes are arrayed along the directive axis rather than the primary axis, which has led some scientists to propose that the bilaterian body was built from a cnidarian-grade ancestor by stretching it out in this directive direction. As bizarre as that sounds, the theory is consistent with certain aspects of early development in some invertebrate bilaterians: the gut often forms from a slitlike blastopore whose middle section zips together, such that the mouth and anus are formed from its two ends. A stretching along the directive axis would also help explain how a system that builds duplicate structures around the blastopore, as happens in cnidarians, might end up deploying such structures along the bilaterian primary axis, although as yet there is no evidence that the developmental photocopying systems of the two groups are comparable.

So, which cnidarian axis was our fore-to-aft axis derived from—primary or directive? It's too early to say, but I have a hunch that the answer may end up being "both." If you remember back to the beginning of this whole story, a newt's organizer sits on the dorsal side of the blastopore, so while its body is drawn out from the back end (consistent with Meinhardt's primary/fore-to-aft axis homology), the extrusion does seem to be focused on one side of the blastopore. If what we're seeing here turns out to be a preserved trace of the origin of the bilaterian body plan, we might therefore conclude that the front end of our cnidarian-grade ancestor gave rise to our front end, while the old back end (i.e., the blastopore end) stretched out and became our trunk, but at an angle, such that the old mouth side is now the ventral side (dorsal in chordates).

However the cnidarian/bilaterian axis homology situation pans out, we have a big problem when considering the transition between the two grades. Whichever way you look at it, it's very difficult to see how a sessile polyp, even one with the seeds of the bilaterian blueprint in its development, could lead to a crawling, flatwormlike animal. But could we be focusing on the wrong part of the cnidarian life cycle? After all, planula larvae are motile and have already proved their worth by helping us to work out which end of the cnidarians is which. Most planulas are very simple, radially symmetrical creatures, but there are some intriguing exceptions. One such oddity is the larva of a species related to *Hydra* called *Clava multicornis*, which has been studied by Stefano Piraino and his colleagues at the University of Salento, Italy. A *Clava* larva is unusually elongate for a planula, and for such a simple organism, it has a remarkably complex nervous system, clearly regionalized along the fore-to-aft axis, with a concentration of sensory cells at the front and longitudinal cords running down its length. Indeed, its level of organization isn't far from that of the simpler bilaterian flatworms. It even has a vague hint of dorsoventral organization—the ectoderm at its front end is slightly thicker on one side. What lifestyle choices has this creature made that could possibly have caused such a departure from the usual planuloid simplicity? Simply put, whereas most planula larvae swim, *Clava* larvae crawl, using the kind of ciliary gliding that may have taken Urbi from place to place.

No one is suggesting that *Clava* larvae are outrageously ancient bilaterian predecessors. Nevertheless, they may be giving us something of an action replay of a change in locomotory technique that ultimately caused the transition from cnidarian- to bilaterian-grade body plans. Most planula larvae, being ciliary swimmers, have no incentive to break their radial symmetry pattern, for the environment is pretty much the same on all sides of their tiny body. When crawling on a surface, however, only the side of the body touching said surface is able to provide thrust, so as long as the animal is consistent in its posture, it will eventually pay to design the contact side along different lines than those of the noncontact side. That doesn't just mean dedicating the ventral surface to locomotion: it might also be a good idea to camouflage or otherwise protect the dorsal surface. Elongating the body in the direction of movement might also be useful, to compensate for the loss of thrust caused by restricting force generation to one side of the body. *Clava* larvae have taken only a tentative step down this path, probably because they don't exist for long enough as planulas for more significant changes to take hold. However, should such a creature become sexually mature before metamorphosing into a polyp, those adjustments might prove worthwhile. Thus might an insignificant crawling larva have ended up siring the great bilaterian radiation.

----------

When it originated, the bilaterian developmental system can have differed only subtly from the cnidarian equivalent: it probably involved just a slight drawing out of the body in the direction of motion. But the consequences were far-reaching indeed. For the cnidarians, the Cambrian explosion was just something that happened to other animals, for one overriding (and of course, largely locomotory) reason. For radially symmetrical creatures, in which structures are repeated *about* the body axis, it's usually of paramount importance that any changes in one module are reliably duplicated in all, as any attempt to make one different will compromise locomotory control. In a bilaterally symmetrical paradigm, however, in which structures are repeated *along* the axis, then as long as left and right copies are tweaked in parallel (the default option), there will be no major disruption to normal locomotory service following a bit of selective tinkering. Hence, there is nothing to curtail a vastly more adventurous exploration of morphological and functional possibilities. As a result, the bilaterians, thanks to their adoption of a crawling lifestyle, were ultimately able to find ways of living and moving in every environment of the biosphere. Ironically, their methods include a whole range of swimming lifestyles—exploited by the arthropods, mollusks, worms, and, of course, the master swimmers, the chordates—that are unavailable to the cnidarian jellyfish, despite the fact that jellyfish are the more ancient occupants of the pelagic realm. Not that that's a problem for these venerable creatures—the cnidarians are a highly successful group of animals, and they are here to stay. But for hundreds of millions of years, they have remained stubbornly conservative in their motions, while their bilaterian cousins went on all kinds of locomotory and morphological escapades. However, none of this would have been possible without what was arguably the most fundamental locomotory transition in the history of our kingdom. An exquisitely organized body is of little use if it lacks the computer to run it. In the next stage of our journey, we will see how this computer—our extraordinary nervous system—came into being.

# 7

# Brain and Brawn

## In which we discover how and why animals became computerized

> I'm playing all the right notes, but not necessarily in the right order.
>
> —Eric Morecambe

It's early spring here in Cambridge, and the weather being particularly pleasant today, I thought I'd give myself a change of scene, so I'm currently writing these words in the University Botanic Garden. The journey was short and simple, but I can confidently state that there exists not a robot on Earth that could have pulled it off. From the very outset, all kinds of obstacles and challenges were thrown my way.

Like many inhabitants of this city, I usually get around by bike. Cycling is a considerably less stable mode of transportation than walking, requiring me to keep my center of mass directly over a strip of ground about an inch wide—the polygon of support under the tires—for much of the time, without the benefit of direct ground contact to provide the odd correcting force should things go wrong. Whenever I turned a corner, the rules changed—now I had to tilt toward the inside of the turn by a precise angle (dependent on the tightness and speed of the turn) to avoid being toppled in the opposite direction by the centrifugal force. My choice of turning speed thus required me to remotely assess the traction I could expect of the wheels—if I were to overestimate this parameter and tear around a bend too fast, I'd heel over so much that the bike would slip out from under me. Being surrounded by momentum-rich cars, my life is likely to have depended on my accurate judgment in this matter.

The first part of my journey was on the main road but, eager to get away from motorized vehicles, I diverted onto a bike path at my earliest convenience, and there faced a whole new set of difficulties: in Cambridge, many of these paths are shared with pedestrians, and this being the weekend, there was a fair multitude present, moving at a variety of speeds in a variety of directions, either of which were subject to change at a moment's notice (especially where children or dogs were concerned). With a sufficiently aggressive bell-ringing strategy I might have been able to clear a straight path, but I wouldn't have made any friends in the process, and in any case, where's the fun in a straight path? So, it was largely up to me to avoid them: a phenomenally complicated computational task involving moment-by-moment plotting of the positions and relative velocities of my fellow path-users, while taking account of their levels of awareness (the smartphone effect) and their potential for chaotic course/speed adjustments (inversely related to age, as a rule), either of which might have required me to give them a wider berth. There were also other cyclists to dodge, which involved a double computation: to avoid them, I had to work out how they were planning on avoiding the pedestrians. The result was a continual flood of complex information that had to be properly processed, then relayed to my muscles to accurately and rapidly adjust my direction and speed, while maintaining stability on what was often a decidedly uneven surface. On at least one occasion I had to deal with a change of surface, when a lack of space forced me off the path onto the surrounding grass—a more deformable, slippery substrate—requiring an instant upping of my power output and a recalibration of my turning strategy.

As I said, it was a mighty complicated task, and that's before considering the fuss of getting on or off the bike without falling over, or how I compensated for the asymmetrical load of a computer hanging off one shoulder, not to mention how I chose a route that would bring me to my desired location, despite my not being able to see the Botanic Garden until I was almost on top of it. Yet here I am. Quite frankly, I'm amazing. I don't mean to boast: you're amazing, too. Day in, day out, whether we use technological enhancements or not, we all pull off truly astounding computational feats of navigation and propulsion that bring us safely and efficiently to our intended destinations. What's more, we rarely give these processes a second thought—most of the countless calculations and decisions involved happen beneath the level of our conscious awareness. If we had to think about each and every muscle contraction (quads on now, pull a bit harder, hamstrings off now, etc.) while registering and implementing the feedback coming in from our senses, our mind would soon collapse under the strain (and our body shortly thereafter) long before we could begin to deal with such higher-order tasks as collision avoidance or working out where we were going.

What all this tells us is that when thinking about the evolution of locomotion, it's not enough to consider only the morphological and mechanical sides of the story. Important as these facets are, it's not just what you've got, it's how you use it that counts, which is where our exquisite nervous system comes into play. This flimsy electronic network of wirelike cells (neurons) must not only direct the generation of effective and efficient propulsive movements—a major task in itself that requires perfect tuning to the layout and properties of the musculoskeletal system—but also gather information from the environment and the body, extract whatever's useful and meaningful, process it, and then deliver an appropriate output to the locomotory machinery to bring about a suitable response that's in harmony with the overall navigational goal. It's a mammoth responsibility, and as we'll see, it was what the nervous system was originally "designed" for. Perhaps more surprisingly, the centralization of a large chunk of nervous tissue into one or more longitudinal cords, with a particularly voluminous collection of nerve cells—a brain—at the front end, was also primarily a locomotory adaptation. We may think of it as, well, the thinking organ—one that allows us to ponder and understand the world around us, empathize with our fellow organisms, and that has somehow granted us consciousness—but all this is a bonus. First and foremost, the brain is for locomotion.

## BREWING UP A STORM

It has long been known that the animal neuromuscular system has an electrical basis. The man largely responsible for this insight was the Italian anatomist Luigi Galvani, who in the 1780s found that freshly amputated frogs' legs twitched when electrically stimulated. The discovery caused a sensation at the time, thanks in no small measure to the ghoulish public demonstrations performed by Galvani's nephew Giovanni Aldini and others, during which the face and body (or body parts) of recently executed criminals were briefly reanimated by the application of an electric current.* However, despite the cultural (as well as literal) shock that this revelation caused, the electrical basis of the neuromuscular system is, from an evolutionary point of view, the easiest of its attributes to account for, as in this respect nerves merely capitalize on properties inherent in all living cells.

Any man-made electrical circuit requires a power source, a conductor (usually a metal wire) to carry the current, and, assuming we want to be able to turn the power on or off, a switch to break or complete the circuit at will. Animal cells are essentially no different, aside from the fact that whereas artificial circuits involve the movement of electrons, biological electricity is based on the movement of ions—atoms that carry either too many or too few negatively charged electrons to balance the positive charge of the nucleus. The main battery chargers of animal cells are pumps (made of protein) within our cell membranes that constantly import and export ions; specifically, potassium ions are drawn into our cells, sodium ions out. Potassium and sodium ions each carry a charge of +1; that is, all are missing a single electron. That should render our sodium/potassium pumps useless as battery chargers, as what we need them to do is accumulate charged particles on one side of the membrane, not just exchange six of one for a half dozen of the other. However, the pumps aren't the only route by which ions can enter or leave our cells. Scattered throughout our cell membranes are a variety of proteins that form ion channels. Many are closed in their neutral configuration, but some are permanently open. A particularly numerous type of channel in this latter category is specific to potassium ions, with the result that our cell membranes, while good at restraining the passive movement of sodium, allow potassium to leak through. Because the concentration of potassium ions is much higher inside our cells than out (thanks to the hard work of the pumps), there is a net emigration of potassium through these leak channels. As there's no equivalent immigration of sodium, the membrane becomes negatively charged on the inside, to the tune of a few tens of millivolts.

In brief, then, our sodium/potassium pumps are effective cell chargers thanks to the much greater permeability of our membranes to potassium than to sodium. If, however, the permeability to sodium were to increase, we would expect the charge to dissipate or indeed reverse, as sodium ions

* Such displays were undoubtedly a major source of inspiration for Mary Shelley's *Frankenstein*, written in 1818.

would be strongly drawn into the cell, partly by the cell's negative charge and partly by the sodium ion concentration difference (remember, the pumps kick sodium out of our cells). We thus have at our disposal a ready means of throwing the switch of an animal circuit: if sodium channels are embedded in the cell membrane, an ionic current will flow whenever the channels are opened. This is how most of our sense organs work, by coupling some stimulus—be it mechanical, chemical, or visual—to the opening of sodium channels. So, we have the batteries, we have the switches—all we now need are some biological wires and we'll have all the elements of a functioning electrical circuit. In one sense this couldn't be easier—carrying current from one part of the body to another merely requires a cell to grow one or more long extensions, which neurons do (the extensions are called axons): the positively charged ions inside will happily carry an electrical current as do electrons in a metal wire. The entry of sodium ions at the sensory end will repel any positive ions in the vicinity (mainly potassium); the repelled ions then repel their positively charged neighbors, the repulsion continues down the length of the axon, and before you know it, the nerve impulse (or as it's known in the business, the wave of depolarization) arrives at the far end.

The use of positive ions to carry an electrical current is not the only difference between biological and man-made circuitry. A functioning artificial circuit must be in complete continuity—break the circuit and the current stops, as anyone with experience of malfunctioning Christmas tree lights knows full well. Neurons, however, only rarely make electrical contact with one another, and they never make electrical contact with muscles. Rather, two cells in a biological circuit are usually separated by an extremely narrow gap called a synapse.† When a nerve impulse arrives at this gap, a chemical called a neurotransmitter is released from the end of the upstream, presynaptic neuron, which then diffuses the short distance to the membrane of the downstream, postsynaptic neuron. This membrane is studded with ion channels that open whenever a neurotransmitter molecule binds. If the ion channel in question is specific for sodium, the binding of neurotransmitter molecules and subsequent influx of sodium ions depolarizes the membrane, and the current heads off down the postsynaptic neuron, as if the two cells had been electrically connected.

---

†Current flow in a single neuron still requires a complete circuit: the movement of ions inside axons would not be possible were it not for the movement of ions in the opposite direction in the fluid bathing the cells.

It may seem odd to break a circuit in this way, particularly as all the business involved with the release and diffusion of neurotransmitter does hold things up a bit. However, synapses are key to the computational sophistication of nervous systems. First, they allow a signal to be amplified, for a presynaptic terminal can be packed with large quantities of neurotransmitter and the postsynaptic membrane similarly packed with ion channels. This property is particularly critical at a nerve/muscle synapse, or neuromuscular junction, because muscle cells—which have to be depolarized to contract—tend to be vastly larger than the neurons that activate them. That, however, is only the tip of the iceberg of computational possibilities. The opening of an ion channel with a neurotransmitter molecule won't necessarily cause the postsynaptic membrane to *de*polarize. Opening a channel (with a different neurotransmitter) that's specific for potassium rather than sodium, for instance, will allow more potassium ions out than usual and thereby *hyper*polarize the membrane—increase the existing negative charge. Such *inhibitory* synapses, as opposed to the more intuitive *excitatory* synapses that I described earlier, are essential to the ability of nervous systems (including brains) to make decisions. A given postsynaptic neuron may receive synaptic input from very many upstream neurons. Some of those synapses will be excitatory, some inhibitory, and some will have a greater intrinsic amplification potential than others. Furthermore, some of those presynaptic neurons may have been more strongly stimulated than others by input from a sense organ. Whether and to what extent the postsynaptic neuron activates depends on the overall balance between excitatory and inhibitory inputs at a given time—there must be significant net depolarization at a synaptic nexus for the signal to continue. Ultimately, the myriad synaptic "decisions" of the nervous system will be reflected in which muscles are turned on in any given moment, and which are turned off—in other words, the behavior of the animal. Even better, the degree to which a given input is amplified at a synapse—and thus the authority it has over its postsynaptic neuron—can be adjusted over the course of an animal's life, by changing the size of a synapse or the density of ion channels in the postsynaptic membrane. This is the cellular basis of learning.

Aside from their role in decision-making, inhibitory synapses are vital for the effective execution of movements. Consider a simple sensory/motor circuit, such as our famous knee-jerk reflex, whereby the quads contract in response to an unauthorized stretch of the patellar tendon to prevent buckling of the leg. The stretch is registered by a sense organ embedded in the tendon called a proprioceptor, one of many that invest our muscles and joints to give the brain and spinal cord detailed moment-by-moment information about the disposition of our body parts (our proprioceptors are what allow us to do such things as bring the tips of our fingers together without looking at them). When the patellar tendon's proprioceptor is activated, it sends nerve impulses via a sensory neuron to the spinal cord, where it synapses with a collection of motor neurons. These motor neurons relay the signal back out to the quad muscles, which promptly contract to extend the knee. So far, so good. But for the reflex to work properly, the quads' antagonists—the knee flexor muscles—must be shut down. To this end, the sensory neuron of the patellar tendon's proprioceptor carries branches that make inhibitory synapses with the motor neurons of the knee flexors (via relaying interneurons). This arrangement ensures that the flexors won't fight the quads during a proprioceptive emergency. As you may have guessed, inhibitory synapses also have the key task of overriding our reflexes whenever our proprioceptors are activated by intentional movements. I wouldn't have been able to sit down to write this if my quads relentlessly straightened my knees every time I tried to bend them.

## THE MOVEMENT COMPUTER

There is one final role of inhibitory synapses that's especially pertinent to our story: the generation of the rhythmic, repetitive movements of locomotion. Such rhythms require coordinated, pacemaker-like output from an animal's motor neurons. Originally, it was thought that chains of proprioceptive reflexes were entirely responsible for such patterns, but we now know that the basic rhythms of locomotion can be generated by networks of neurons in the central nervous system in the complete absence of sensory input. This fact was first discovered by neurobiologist Donald Wilson in the 1960s: he found that the motor neurons of the flight muscles of a locust would continue to produce a broadly normal output (if on the slow side) following decapitation, removal of the abdomen, and, most significantly, amputation of the wings. Without the wings, the central nervous system was robbed of the rhythmic proprioceptive feedback that had been thought essential to the generation of the repetitive motor pattern.

It would take many years of painstaking work to piece together the neural circuitry of the locust's so-called central pattern generators (CPGs), but in essence the system is quite simple. Theoretically, two neurons can do it. Say neuron A makes an excitatory synaptic connection to neuron B, while neuron B makes inhibitory contact with A (Fig. 7-1). Whenever A is turned on by some upstream nervous input (from the brain, for instance), it turns B on, and thereby turns itself off again shortly afterward thanks to the inhibitory connection between B and A. Without the signal from A, however, B falls silent; so as long as the original input from the brain is still there, A will start up again. Of course, as soon as it does so, B kicks into action once more, A then shuts up, and so on and so on. The logic is the same as that used in the Notch/Delta segmentation clock, with inhibitory synapses replacing self-inhibiting proteins, and it gives exactly the kind of rhythmic pulsing that the locomotory machinery requires, with a frequency that depends on the delays involved in the transmission of electrical signals between the two neurons. In reality, the locust flight central pattern generator is a bit more complicated than this—for one, there are actually several self-inhibiting circuits arranged in parallel, presumably to make the system more robust in the face of potential perturbations—but the basic principle holds.

Since their discovery in locusts, CPGs have been found in a wide range of animals, from sea slugs to cats, and there's now little doubt that they underlie rhythmic locomotory muscular contractions across the kingdom. The nervous system of a given animal usually contains several such circuits—one set for each oscillating unit (e.g., a leg)—whose individual rhythms are coordinated as necessary, usually metachronally. All are capable of running without sensory input. However, just because these movements *can* be generated in the absence of sensory feedback, that doesn't mean that such feedback isn't necessary for effective locomotion. If the system blithely chugged away without paying any attention to the proprioceptors, it would take only the slightest irregularity to knock an animal over. It's therefore crucial that the proprioceptors talk to and modulate the activity of the CPGs. When we walk, for

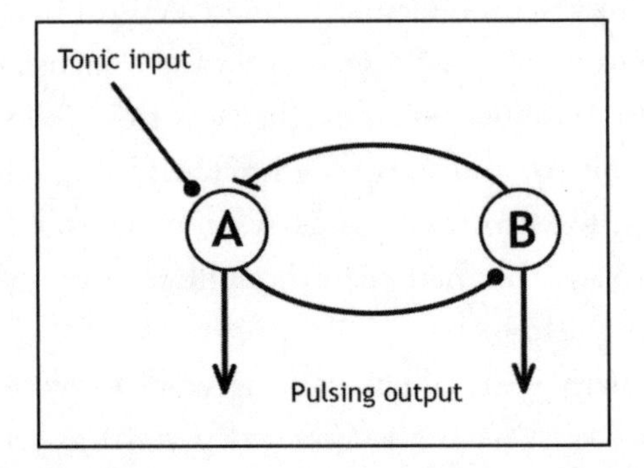


7-1: A simple central pattern generator: A excites B and B inhibits A, with A receiving continuous input either directly from a sense organ or via the brain.

instance, the leg proprioceptors must signal that the leg is not supporting any weight before its swing phase is initiated, otherwise we'd collapse.

Of course, it's also essential that the motor output can be modulated by navigational commands mediated by the long-range senses of vision, smell, hearing, and the like, so that an animal changes course and speed when it needs to. Also important, particularly for big terrestrial animals like ourselves, are commands relating to the sense of balance, mediated in vertebrates by the mechanical receptors of the inner ears. These myriad inputs must be correctly processed and prioritized by the nervous system (all done by appropriately weighted excitatory and inhibitory synaptic inputs) to give the optimum response from the musculoskeletal system. In a nutshell, that is the brain's raison d'être. Learning, memory, object recognition, problem-solving, even emotions and consciousness—when all's said and done, these are just particularly elaborate ways of sifting through inputs and generating appropriate muscular outputs.

You might be alarmed by this apparently prosaic description of what are arguably the most astonishing objects in the universe. But to us evolutionary biologists, the simple foundations of our wondrous brain, while providing a poetic beauty of their own in my view, constitute a fantastically useful explanatory olive branch that gives us the means to understand how our brain came to be. However, while it is true that the function of the nervous system is based on a mere collection of ion pumps and channels, together with the use of excitatory and inhibitory neurotransmitters that enable neurons to evaluate inputs and produce appropriately calibrated and patterned outputs, our explanatory task is at yet barely begun. If we are to understand the evolutionary origin of the nervous system and brain, we must do far more than point out the simplicity and ancient pedigree of the components. The whole is far, far greater than the sum of its parts, and the assembly of the whole is what we need to account for. Somehow, from a simple, prenervous starting point, evolution gradually brought the labyrinthine locomotory computer into existence. And we're not talking about any old elaborate structure here. The evolution of that quintessentially complex organ—the human eye—can actually be accounted for quite easily. Indeed, computer simulations can build virtual eyes from scratch, using the rules of natural selection, in a few hundred thousand generations: a blink of an eye in geological terms. But when it comes to the nervous system and the way it interacts with the muscular system that it controls, we're dealing with a different sort of problem. The eye, for all its intricacy, is a fairly self-contained structure, and has been since it was a mere whippersnapper of a patch of light-sensitive cells, whereas the neuromuscular system is a vast, intricately connected network whose main cell types—nerves and muscles—are derived from entirely different germ layers—ectoderm and (usually) mesoderm, respectively. Yet the component cells somehow managed to connect to one another in the most bewilderingly complicated ways. Add to that the troubling observation that neither nerves nor muscles seem to be of any use without their partner tissue, raising the old "which came first?" question. How did it all start?

## SNEEZING ON THE SEABED

When hunting around for clues regarding the origin of our brains, sponges are probably the last creatures you would expect to look to. Among the simplest animals on the planet, they have no nerves, no muscles, and no sense organs, and seem to be built for one thing only: filtering microorganisms—mainly bacteria—from seawater. Indeed, a casual observer would be hard-pressed to classify them as animals at all (Aristotle certainly agonized over this point). And yet, the rest of our mighty kingdom may well have sprung from such as they. Recent analyses of sponge relationships have revealed that they may not be a natural group, but rather a series of ancient evolutionary divergences from the bowels of the animal family tree. That's great news if it's true, because it would mean that the last common ancestor of the sponges, which must have had all the features shared by the modern groups, was also *our* ancestor. We would therefore have an unusually clear picture of what we were like back then, by which I mean anything between 770 and 850 million years ago. If we're looking for a snapshot of a prenervous animal, sponges are therefore as close to perfect as we're likely to get.

But what can these apparently motionless, filtering bags actually tell us, other than demonstrate just how little an animal can do without nerves and muscles? Well, there's more to sponges than meets the eye. It's widely believed that the only bits of them that move are their beating cilia that waft water through their tissues, but that is far from the truth. Surprising as it may seem, many of them can actively contract their "body" after a fashion. This has the effect of speeding up the expulsion of water, rather like a slow-motion sneeze, which presumably stops the filtration gear from getting clogged up with sediment. Consistent with this idea, the contractions are stimulated by increased quantities of sediment in the water. What's more, the squeezes often happen in surprisingly regular chains, although the frequencies are so slow (two beats per hour or slower) that this fact is usually

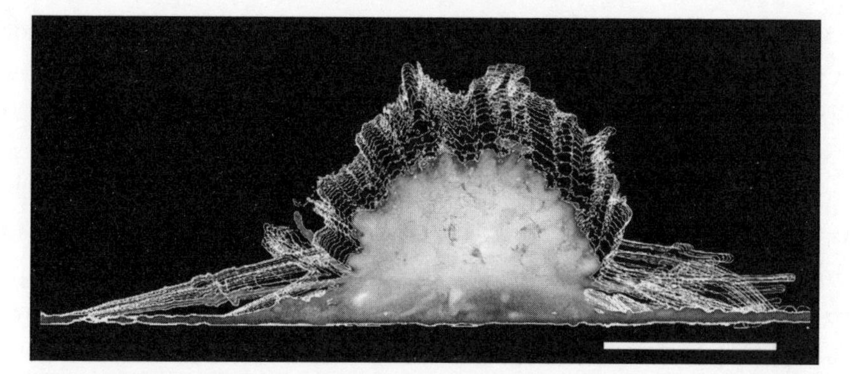

7-2: The contraction range of the sponge *Tethya wilhelma*. The movement shown occurs over forty-five minutes. Scale bar = 5 millimeters.

overlooked. Sponges only reveal their true colors if time-lapse photography is used to compress several hours of their lives into seconds. In this accelerated light, they pulse like malformed jellyfish.

The whole-body coordination of sponge contractions, their regularity, and their responsiveness to environmental stimuli are attracting a lot of research attention at the moment, for these properties would usually be underpinned by a neuromuscular system, which of course sponges lack. While we don't yet have a complete understanding of how they do it, the pieces of the jigsaw puzzle are steadily falling into place. One such piece has been provided by Michael Nickel of Friedrich-Schiller University in Jena, Germany. He looked carefully at the way a sponge's tissues change size and shape during the contraction cycle, and found that the biggest changes occur in a sponge's outer cell layer, which lines not only the nominal external surface of the creature but also a system of canals and chambers that extend internally. During a contraction, the surface area of this layer decreases almost by a factor of three, suggesting that its cells house the contractile machinery. The genes that make important parts of the molecular apparatus of muscle have been found in sponges and are strongly expressed in the layer concerned, lending support to this idea. The cells don't exactly look the part, being wide and flat, but with such ultraslow pulse frequencies, there's really no need for any kind of bulk: a light smattering of contractile molecules should be quite sufficient for their needs.

Rather more perplexing than the business end of sponge contractions are their responsiveness and coordination. Here we're largely in the dark (and how wonderful it is in this day and age to be able to say that about the basic function of a major group of animals!). That said, Sally Leys and her lab at the University of Alberta have found that sponge contractions can be induced by certain chemicals, including glutamate (of monosodium glutamate fame). Indeed, in one experiment, a high concentration of glutamate triggered such vigorous contractions (by sponge standards) that the sponge in question tore a section of its body apart. Conversely, a second chemical—gamma-amino-butyric acid (GABA)—was found to inhibit contractions in one species (although it stimulated them in another). Suggestively, both glutamate and GABA are used as neurotransmitters in more complex animals, GABA at inhibitory synapses. Leys and her colleagues have also found what appear to be rudimentary sensory cells that carry modified, nonbeating cilia. These are dotted around a sponge's main exit flues, and must play a role in stimulating contractions, for if the cilia are removed by treating the sponge with chloral hydrate, the contractions stop.

Putting everything together, it now seems that sponges have many of the basic components of a functioning neuromuscular system: contractile cells that respond to applied chemicals as muscles respond to neurotransmitters, and sensory cells that presumably release such chemicals when stimulated. The only bits that sponges are missing are the nerves themselves and the attendant electrical signaling.* The likelihood is that the protoneurotransmitter chemicals (if I may call them that) are broadcast throughout the body of the sponge in much the same way that hormones are broadcast in more complex animals. That makes the signaling extremely slow by nervous standards, but with a required contraction rate of only two per hour, there is manifestly no need for speedy intercellular communication.

## CHILD'S PLAY

Thanks to sponges, we now have an idea, albeit tentative and hazy, about how the neuromuscular system got started. But what of the next critical step—the origin of the all-important nerves? If we follow our evolutionary lineage up from the last common ancestor we share with sponges, the next major divergence we come to (at least, the next one that nearly everyone agrees on†) is

* The beautiful glass sponges are the exception to this general rule, but that's because the cells form a huge, interconnected network, as if the whole thing were a single cell. Their unusually rigid construction rules out contractions, and electrical signals instead shut down the filtering current.

† Two groups whose placements on the animal family tree cause perennial trouble are the vaguely jellyfish-like comb jellies and the even more obscure placozoans, which resemble nothing so much as giant multicellular amoebas. Recent analyses put the comb jellies basal even to the sponges, but it remains to be seen whether this controversial result stands the test of time.

between our own bilaterian group and the cnidarians—the anemones, corals, jellyfish, and whatnot. While we're often tempted to regard these creatures as basic and primitive, they do all have fully realized nervous systems (which reach impressive levels of sophistication in one group). The all-important transitional forms between the sponge grade and the cnidarian/bilaterian last common ancestor have unfortunately left no trace. Or have they? While sponges and cnidarians are poles apart as adults, their larvae have much in common. Both kinds swim using cilia, which cover their tiny bodies like wriggling fur coats, and both have basic navigational skills by which they find their way to a suitable spot to settle. Significantly, many bilaterians have simple ciliated larvae, too. But whereas sponge larvae make do without any internal wiring, most of their bilaterian and cnidarian counterparts have rudimentary nervous systems. Could such creatures give us the next big piece of the neuromuscular jigsaw puzzle? It's a tantalizing prospect, with just one snag. These larvae are tiny—no more than a few hundredths of an inch across. How on Earth can the function of the nervous systems of such minuscule animals be teased apart?

Gáspár Jékely, based at the Max Planck Institute for Developmental Biology in Tübingen, was undaunted by this most daunting of challenges. In the last few years he and his research group have carried out some quite stunning work to shed light on the locomotory computer of a larval worm: a marine annelid called *Platynereis dumerilii*. Freshly hatched, this creature is nearly spherical—roughly one sixth of a millimeter (seven thousandths of an inch) across—with a band of cilia running around its middle and, just in front, a pair of simple eyespots that each consist of two cells: the photoreceptor itself and a neighboring screening cell packed with dark pigment, which ensures that the receptor is only stimulated by light coming from one side. The larvae use this basic equipment to swim toward the light—they are, as we say, positively phototactic—which in the wild is thought to ensure that they stay up in the water until they're ready to settle down (as adults they live in self-constructed tubes). Interestingly, they don't move in straight lines, but follow curious spiraling trajectories, with the dorsal side of their tiny body always directed toward the axis of the spiral.

It was the nervous control of this light-guided movement that Jékely sought to understand. Conventional dissection techniques being obviously impractical for something this small, he turned instead to the long-established method of serial sectioning. This involves slicing the object of interest into a set of thin layers—once each layer is photographed, the whole structure can

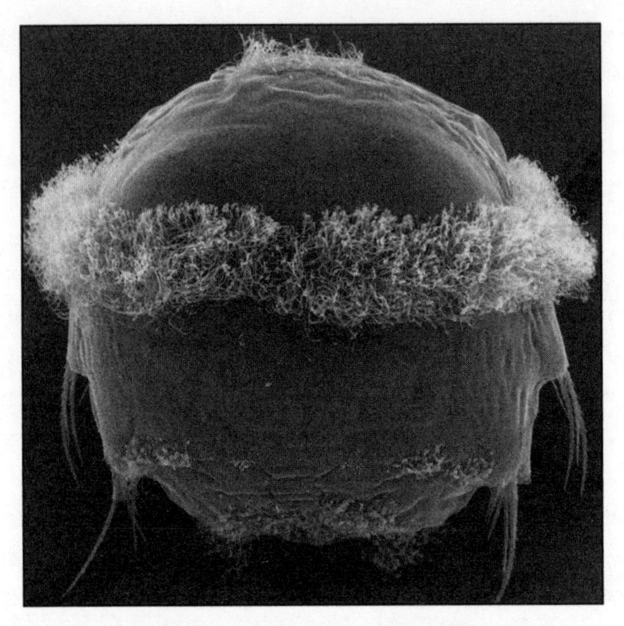

7-3: The larva of the annelid worm *Platynereis dumerilii.*

be digitally reassembled by aligning the 2-D images. Of course, for an object that's one sixth of a millimeter in diameter, those layers have to be *very* thin—about 70 nanometers* across—so the slicing had to be done under an electron microscope. With such high resolution, all sorts of juicy details, such as the disposition of the synapses, could be precisely mapped. Jékely thus found that each photoreceptor doubles up as a motor neuron, whose axon makes synaptic contact with the nearest ciliated cells at the far left or right of the band (whichever side is closest). If an eyespot of a living larva was illuminated by a narrow beam of light, the signal from the photoreceptor caused the lateral cilia on the stimulated side to reduce their beat frequency.

This simple light-responsive mechanism neatly explains *Platynereis*'s phototactic behavior. As the larva corkscrews through the water, its eyespots give regular updates on the ambient light levels. If the light is consistently brighter on one side, the periodic slowing down of the lateral cilia on the more brightly illuminated side gradually steers the creature toward the light. The spiral path ensures that the photoreceptors don't adapt to the ambient light gradient (their eyes, like ours, rapidly accommodate to the general level of illumination), which would cause them to stop steering prematurely and wind up on the wrong heading. The twirling is brought about by slight

* One nanometer is one billionth of a meter.

asymmetries in the operation of the ciliary band: the dorsal cilia create a slower flow of water than their ventral counterparts and all beat along a slight diagonal, slewing the wake to one side.†

The cilia-driven, spiral phototaxis technique is widely used by small aquatic creatures. Notably, sponge larvae do it, too, even though they don't have eyespots like *Platynereis*. Instead, they've turned some of their ciliated cells (they have plenty to spare) into multipurpose cilia-bearing photoreceptors, easily recognizable by their accumulation of screening pigment.‡ Because light detection is an expensive business, involving the continuous manufacture or regeneration of pigment molecules, costs are often cut in these nerveless larvae by scrapping the beating movements in the photoreceptive cells: the downgraded cilia instead act as rigid rudders. More surprisingly, despite its comparative simplicity, this nerve-free phototactic strategy is just as effective at light-chasing as the wired-up version shown in *Platynereis*.

So, and note the significance of this question, why develop nerves? As is so often the case in the evolution of locomotion, the answer likely boils down to simple economics. Jékely thinks that the main benefit of adding nerves is an improvement in what he calls the sensory-to-motor transformation. For a larva to alter its trajectory without nerves, it requires many expensive multipurpose sensory/steering cells—roughly one tenth of a sponge larva's surface cells are of this kind. But if a given sensory cell can affect many ciliated motor cells via axons and amplifying synapses, one or two detector cells may be enough. The transformation could have been an easy one. If a sensory/rudder cell were to extend its influence by secreting a chemical that altered the beating frequency of its ciliated neighbors—not a million miles away from what the sensory cells of sponge adults likely do—it would effectively have become a combined sensory/motor protoneuron (and the chemical a protoneurotransmitter). From here the cell would have needed only to extend protuberances to those same neighbors to take the next step toward full-blown neuronal status. At some point, electrical conduction would have had to step in to rapidly couple the stimulus-detecting events at one end of the cell to the release

---

†Jékely's group characterized the system so completely that they've been able to build a hyperaccurate interactive simulation of *Platynereis* phototaxis, available at www.cytosim.org/platynereis. Have a play; it's surprisingly therapeutic!

‡In the larvae of the sponge species *Amphimedon queenslandica*, whose genetics are particularly well known, the pigmented cells form a ring near the back end (as defined with respect to the direction of motion). The differentiation of these cells is initiated, at least in part, by a Wnt morphogen that, as I said in Chapter 6, is concentrated at the rear.

of protoneurotransmitter at the other, but as we've already seen, much of the necessary molecular gear was in all probability already in use. With the further growth of the protoaxons, more and more motor cells would have come under the sensory cell's influence, until eventually it reached the not-too dizzying heights of a *Platynereis*-type photoreceptor.

That, however, was just the beginning. The reduced costs of the neuronal, as opposed to the nonneuronal arrangement, opened the door to greater specialization of the sensory cells and the invention of new kinds of detector apparatus. Of course, the increasingly diverse sensory inputs would need to be evaluated and prioritized, so any synaptic contact between individual sensory/motor neurons would have been strongly selected for. Furthermore, if one of the sensory/motor neurons were to slip beneath the surface cell layer, it could thereafter ditch its sensory functions and concentrate on gathering information from other sensory cells—in other words, we would see the beginnings of a division of labor in the originally multifunctional neuron population into dedicated sets of sensory and motor neurons. With the addition of more links to the sensory/motor chain—in other words, interneurons—our little ancestor would have been able to carry out increasingly elaborate computations. Thus a simple brain was born.

## MUSCLING IN ON THE ACTION

The scenario just painted—that nerves evolved in larvae to control ciliary steering, following the evolutionary divergence of sponges—implies that we should find nervous/ciliary circuits in the planula larvae of cnidarians. Unfortunately, while most planulas are known to possess rudimentary nervous systems, we know very little about how their nerves are involved in ciliary locomotion,* although it would be extremely surprising if they were found to have no locomotory role at all. We're just going to have to watch this space, I'm afraid. At present, however, we have an elephant in the room to deal with. A nervous hijacking of cilia is one thing, but our main concern is how nerves ended up in control of muscles. Can our rummaging in the depths of the animal kingdom throw up any useful information about this most important of transitions?

---

* If you're thinking, "What else *could* planula nerves be doing?" it's fairly well established that at least part of the larval nervous system initiates metamorphosis—the larva's transformation into a polyp—when the time and place are right.

We already have one key piece of evidence at our disposal—the contractile cell layer of sponges. Is there anything comparable in cnidarians? As it happens, yes, there is. We bilaterians, as noted in the last chapter, derive our tissues from three germ layers—and our muscles are largely built from the middle, mesoderm layer. Cnidarians, however, lack mesoderm, so their musculature is instead built into the ectoderm, which forms their outer surface, and endoderm, which lines the gut. From one side (the outside if we're talking about the ectoderm, the inside of the gut if we're talking about the endoderm), the cells appear to form typical, tessellated layers, like our epidermal cells. Where the cnidarian layers differ from the sponge version, however, is that underneath, each cell broadens at its base, and is often drawn out into a long tail that runs parallel to the surface. These broad bases and tails contain the bundles of molecular fibers that bring about contraction. With their denser aggregations of contractile molecules, cnidarian muscles, while considerably weaker than the dedicated mesodermal muscle cells of bilaterians, are much stronger than the equivalent cells of sponges, and so can bring about much faster changes of body shape.

Of course, fast contraction is of little use without comparably rapid signaling—the untargeted release of protoneurotransmitters by sensory cells, as presumably happens in sponges, just won't do. Accordingly, even the simplest cnidarian polyps have diffuse networks of nerve cells to coordinate the contraction of their muscles. Where these nerves came from is still unknown, although there are two leading theories. First, there may have been a replay of what happened to the ciliary system, with originally nonnervous spongelike sensory cells growing axons toward contractile cells in the vicinity to improve the speed and efficiency of sensory-to-motor transformation. Alternatively, the older, larval, ciliary nervous system may have taken on the directorship of the developing musculature. Indeed, there is a distinct possibility that both theories could be partially correct, meaning that what we think of as a single nervous system is actually a chimera—an assemblage of formerly distinct populations of nerve cells with different evolutionary origins.

However the wiring of the muscles occurred, the extra control afforded by the nerve nets unlocked opportunities that went far beyond an increased speed of communication between sensory and motor cells. The behavioral repertoire of sponges is limited to a crude, wholesale contraction of the entire body. Once nerves were in place, however, it became possible to selectively activate distinct portions of the musculature to bring about more nuanced movements. The body could be bent from side to side, or, with the segregation of the muscle cells into a set with longitudinally aligned tails and an antagonistic set with circularly aligned tails, the body could be alternately elongated and shortened as in worms. Independently operable tentacles could be grown. In short, with the nervous control of muscles came recognizably animal behaviors.

That said, it's very likely that these early neuromuscular developments occurred in a nonlocomotory context. The primitive life cycle of the cnidarians probably involved only a ciliary planula larva and a sessile polyp adult: this is what we see in the anthozoans, which are the most primitive living group of cnidarians. That makes sense—getting to grips with the musculature is much easier if you don't initially have to worry about getting from A to B (kind of like learning to fly in a simulator before being let loose on the real thing). However, the propulsive potential of their new neuromuscular equipment was not lost on all cnidarians. While sea anemones are popularly regarded as entirely static organisms, many can use muscle contractions of the body column to inch from place to place. Some remarkable anemones can even swim, after a fashion, either by thrashing their tentacles up and down or, in members of the *Stomphia* genus, by inflating the tentacle end of the body into a broad disk and swinging the body from side to side. They do this to escape from starfish (albeit slowly). But such crude locomotory experiments pale in comparison with the better-known examples of the co-option of the cnidarian neuromuscular system for locomotion.

## THE GELATINOUS JET SET

As explained in the last chapter, while anthozoans make do with a simple two-phase life cycle, the more advanced hydrozoans and scyphozoans have added a third phase—the medusa, or as it's more popularly known, the jellyfish (although some rather fickle hydrozoans have since suppressed this phase). Although there doesn't appear to be much to jellyfish—each is little more than a hemispherical, pulsating swimming bell with a bit of buoyant jelly and a set of trailing tentacles—jellyfish represent the evolutionary culmination of a far more concerted attempt to exploit the locomotory potential of the neuromuscular system than is seen in any anthozoan. Their technique is usually described as jet propulsion, with the swimming bell alternately drawing in and expelling water, although, strictly speaking, this is only one of two swimming strategies used by the group. Most scyphozoan medusae (or scyphomedusae, for short) and some hydromedusae are too flat to make use of jet propulsion, and instead row through the water, using the edge of their swimming bell as a circular oar. True jetters have deeper, somewhat bullet-shaped bells equipped with an exit nozzle. The two locomotory styles are associated with distinct ways of life. Jet propulsion is fast but costly, so the jellyfish that use it are ambush predators that spend most of their time waiting for an actively swimming animal to fall into their deadly embrace; they only switch on their engine to escape from anything that might want to eat them. The rowers, on the other hand, are cruisers, whose cheap, sedate swimming style is ideal for catching small, drifting plankton.

All jellyfish, be they rowers or jetters, use the same basic muscular system, and operate it in a broadly similar fashion. The underside of the swimming bell is built of the usual cnidarian multipurpose ectodermal cells with muscular tails that we saw in anthozoans. The tails are aligned with the circumference of the jellyfish rather like the circular muscles of bilaterians, and are responsible for squeezing the bell to impart rearward momentum to the water. The necessary antagonistic action is provided by elastic recoil of the jelly. The recoil is relatively slow, but then, it needs to be: if bell expansion were as fast as bell squeezing, the backward momentum gained by the jellyfish as water flowed forward into the bell would equal the forward momentum it gained during the power stroke, and the creature would simply shuttle to and fro. In the scyphozoa, the muscle cells are synchronized with a simple nerve net, but this is rendered partly unnecessary in hydrozoans because their muscle cells are electrically connected, rather like our heart muscle cells, and thus can synchronize themselves. The two groups also differ in how they control the regular pulsing of the swimming muscles. The scyphozoan control circuits are located in clusters at the bell margin called rhopalia. There are usually either eight or sixteen of these, and if they're removed, the jellyfish can't swim. Each rhopalium contains simple eyes, gravity detectors, and pacemaking CPG circuits: they're basically mini-brains that gather information about the local conditions at their segment of the swimming bell and collectively set the contraction pulse. Hydrozoans, on the other hand, lack rhopalia—the entire margin of the swimming bell is instead studded with eyespots, and the control circuitry is concentrated into two nearby parallel nerve rings. Their version seems to work just as well as the scyphozoan system, and some suspect that the fundamentally different organization indicates that the two cnidarian groups evolved their medusas independently.

The apex of cnidarian neuromuscular organization is seen in the notorious box jellyfish. These are now known to be highly modified scyphozoans,

7-4: A box jellyfish (*Carybdea branchi*), giving a clear view of the nozzle at the edge of the swimming bell. There is a rhopalium on each side of the box.

but unlike their rowing kin they use jet propulsion, and are exceptionally maneuverable: they can steer their way through underwater obstacle courses by using a directable nozzle on the box-shaped swimming bell. To support their agility, they have twenty-four eyes, six in each of four rhopalia embedded in the sides of the bell. These are no mere patches of photoreceptors: remarkably, eight (two per rhopalium) are complex, camera-like organs, complete with lenses. Uniquely among the scyphozoa, the rhopalia are linked by a condensed nerve ring. All in all, the box jellyfish system suggests an unusually sophisticated level of integration.

Anna Lisa Stöckl and her colleagues at Lund University, Sweden, recently put box jellyfish coordination to the test, by seeing how they responded to changed lighting conditions perceived by a single rhopalium, while the creature was held in place by a gentle suction force applied to the top of the bell. They found that the CPG frequency of the rhopalium in question went up when the light levels dropped suddenly. In the species' natural mangrove forest habitat, such a dimming would indicate either the proximity of an obstacle or that the jellyfish had just moved into a shadow, where pickings are slim (their tiny crustacean prey congregate in shafts of light). Both situations are bad news, so we can understand why the jellyfish would put pedal to the metal. Stöckl also found that the rhopalium whose CPG was running at the highest frequency would override the others, probably through synaptic inhibition, and dictate the swimming speed. In other words, the box jellyfish's nervous system is set up such that any one of its four "brains" can take the reins if it sees trouble. Presumably the leading rhopalium not only hits the accelerator but also takes command of the nozzle in such situations, but at the time of writing this has yet to be looked into.

## THOUGHTFUL CRAWLERS

Box jellyfish are undoubtedly the most intelligent cnidarians on Earth (not that the competition for the title is particularly stiff). However sophisticated the box jellyfish nervous system is, though, it's still a far cry from what the bilaterians have been able to achieve. This raises an immediate question: why have box jellyfish never condensed their control circuitry into a single brain? It seems the obvious next step.

In the last chapter we met the planula larva of the hydrozoan *Clava multicornis*, which crawls rather than swims from place to place, unlike most self-respecting planulas. Significantly, given our current concerns, it has an unusually complex nervous system, including a dense network of neurons toward its front end, and we might now wonder why. While we don't yet have a definitive answer to this question—the neural complexity of *Clava* larvae was discovered only recently—we can make some educated guesses. First, we might expect effective navigation over a surface to be more computationally demanding than doing the same in the water, in that it requires tactile and chemical substrate assessments that simply don't apply when swimming. Furthermore, the usual spiral trajectory that works so well for swimming planulas is, of course, impossible on a surface, and while *Clava* larvae still use ciliary propulsion (on a bed of mucus), they have to use muscular contractions to steer. The anterior concentration of nerves is almost certainly a reflection of an increased concentration of sensory cells at the front end. While running a locomotory system with nerves is often cheaper than doing so without, neuronal wiring is still costly to maintain, what with all those ion pumps, so there's a strong incentive to keep the overall length of axons to a bare minimum. If a variety of sensory inputs need to be processed, it therefore makes sense to concentrate the relevant circuitry as close to the sensory cells as possible.

I'd now like to run a simple thought experiment. We're going to take a standard swimming planula and a *Clava*-like crawling planula and increase their size (bear in mind that the growth of a crawling planula might well have been how the great bilaterian dynasty got started, as argued in the last chapter). Before long, both of our overgrown planulas will run into trouble. Ciliary propulsion works perfectly well for small creatures, but this engine quickly becomes ineffective as body size increases. As I will show you in Chapter 9, the diameter of a cilium is more or less fixed by its internal molecular architecture, so if you were to increase its length to try to compensate for a more massive body, it would become uselessly floppy. Conversely, if you decided instead to keep the ciliary array close-cropped, the amount of fluid propelled into the wake would be pathetically insignificant, relative to body mass. Beyond a certain size, the creature must therefore turn to muscle power. As our swimmer is a radially symmetrical animal, its best option is to transform into a relatively broad, flat medusa (in reality, of course, this didn't happen directly, but via the polyp).* Our crawling planula, on the other hand, will be best served by drawing itself into an elongated worm, to make use of the various crawling strategies outlined in the last chapter. As it happens, *Clava* larvae have already taken a step down this road—they are unusually long and thin, by typical planula standards.

This simple difference in the optimal shape of a muscular crawler versus that of a radially symmetrical muscular swimmer had enormous implications for the evolutionary future of the nervous system. In a medusa, important sensory information can come from all sides, so with the high cost of wiring, it's just too expensive to appoint a single brain—the processing circuitry would be too distant from the chief sensory inputs. What we see in a box jellyfish is a nervous system that, for a medusa-style body, is as centralized as it wants to get. The four connected rhopalia, with their power-sharing protocol, provide the best compromise between the dense circuitry required for elaborate processing and the distributed circuitry that matches the necessarily scattered nature of its senses. In an elongated, wormlike body, no such compromise exists. The bulk of the important sensory information reliably comes from one end, so there's no economic reason why there shouldn't be a single concentration of neural circuitry there to process the inputs, and relatively little to stop the evolution of ever-more-complex interneuronal connections within the processor.

There is another more subtle but no less significant difference between medusa-like swimmers and wormlike crawlers, in terms of how best to run

* Were it bilaterally symmetrical, it could adopt a more fishlike body and swimming style, but you have to have been a crawler at some point in your evolutionary past to have such a symmetry system.

a muscular propulsion system. In the last chapter I drew attention to the near-universal use of Mexican waves by the bilaterians—an operational regime that gives great locomotory efficiency, thanks to the smooth, consistent thrust it provides. Ciliary systems, as seen in sponge, cnidarian, and bilaterian larvae, are run in the same way, with predictable results: computer simulations of cilia arrays carried out by Jens Elgeti and Gerhard Gompper of the Jülich Research Centre, Germany, indicate that synchronous beats give only one tenth the efficiency of a well-timed metachronal wave. What might be more surprising is that ciliary metachronal waves happen entirely automatically—they require no nervous coordination, hence my mention of nerveless sponge larvae a couple of sentences ago (we'll see how cilia self-coordinate in Chapter 9). But, as we've seen, cilia don't work if you're big, and unfortunately for medusae, there is simply no way of implementing Mexican waves, given their propulsive strategy—their locomotion is necessarily stop/start. That does mean, however, that they can get away with a very simple motor nerve system, as they will always want to contract the entire propulsive musculature of the swimming bell at once.

In the bilaterians, however, the wisdom of the old ciliary system wasn't forgotten, for there's no reason why an elongated, overgrown planula that would rather crawl than swim can't use metachronal waves. But unlike ciliary arrays, muscle-based systems don't automatically produce such waves—the pattern has to be elicited nervously. This challenge gives us a different perspective on the defining aspects of bilaterian development that we looked at in the last chapter: the posterior extrusion (the proverbial pulling of our head out of our anus) and the segmentation clock, which seems to have been primitively involved in building the repetitive, ladderlike pattern of the bilaterian nervous system. Such a pattern is just the ticket if you're trying to activate transverse muscle bands sequentially. We've already seen how such a system unlocked a cornucopia of complex locomotory technologies in the bilaterians. That anatomical complexity inevitably led to a corresponding increase in the complexity of the brains that controlled them, and the rest is history. When all's said and done, our wondrous mind might owe its existence to an overgrown larva's attempt to hang on to the metachronal magnificence of its ciliary forebears; and the fact that you and I can actually understand all this is really just another expression of this ancient winning design.

# 8

# Give It a Rest

## In which we learn that even the laziest organisms aren't immune to life's wanderlust

For God's sake, let us sit upon the ground.

—William Shakespeare, *Richard II*

All the signs are right. The lustrous red pigment with its intense ultraviolet sheen is just what the fly's been looking for. The rich, fragrant aroma is irresistible. And, in any case, it's a plant—what's the worst that could happen? So, the insect touches down, fully expecting to gorge itself on a banquet of overripe fruit. Sadly, an altogether more grisly fate awaits it. As the fly wanders around, lapping at the droplets of nectar that ooze from the edges of its landing platform, it brushes against a stiff hair—one of six that stud the surface. Nothing happens, and the fly carries on. But a couple of seconds later it touches another, and from that moment it's doomed. The second touch triggered an electrical signal that now spreads rapidly across what is in fact a modified leaf, causing its two halves to snap shut. The unfortunate fly is now imprisoned, its desperate struggles serving only to further stimulate the trap, which responds by clamping shut ever more tightly while releasing a deadly cocktail of digestive enzymes that slowly break the insect's body down into a nutritious soup. A few days later, there's nothing left but a few fragments of exoskeleton—the plant has absorbed everything else.

I have of course just described an ambush by a Venus flytrap. This has to be one of the best-known plant species in the world, despite the fact that few people ever come across it in the wild: it grows only in wetlands within a 60-mile radius of Wilmington, North Carolina. Its notoriety, undiluted by its restricted geographical range, owes much to its ability to turn the ecological tables on animals, although it's far from unique in its eating habits: worldwide, there are about six hundred species of carnivorous plants. What really sets the Venus flytrap apart—the thing that caused Darwin to call it one of the most wonderful plants in the world—is the *way* in which it captures animals. It doesn't just passively entice insects into a pool of water, like pitcher plants, or trap them in glue, like sundews: it literally grabs them, using the kind of rapid, reversible movements, complete with electrical activation, that are usually seen only in animals. Once witnessed, a trick so unique and unexpected can never be forgotten.

The Venus flytrap's behavior is a startling jolt to the comfortable "animals move, plants don't" dichotomy that we all take for granted. It's the exception that proves the rule, causing us to consider a question we rarely ask: *why* are the two kingdoms so different in this regard? Locomotion has completely dominated the evolutionary history of the animal kingdom, to the extent that it's difficult to find any aspect of an animal's anatomy that hasn't been touched either directly or indirectly by the need to move. How could something that is for us so important be so singularly irrelevant for our plant cousins? Even the Venus flytrap—the most energetic of plants—never uses its skills to wander about.

Of course, plants might ask us animals the same question, and frankly, they'd have a point. In many ways, ours is the paradoxical strategy. Why do we insist on devoting so much energy to tearing around, when we might otherwise invest those resources in reproduction—the only thing that ultimately matters? Even our obsession with locomotory efficiency in its countless manifestations, from the Achilles tendon to Mexican waves, can only reduce the bill so far. Foraging bumblebees use up an extraordinary 80 percent of their daily energy budget on flight. Admittedly, that's an extreme example, but trout expend as much as 30 percent of their available funds on locomotion, lizards burn about 20 percent, and chimpanzees, up to 50 percent, if the exceptional costs of warm-bloodedness are excluded from the balance sheet. And all that considers only the direct costs of locomotion.

What of the hidden costs of building and maintaining the locomotory apparatus and its support tissues, such as the skeleton and nervous system? For us humans, just keeping the brain going consumes about 20 percent of our daily fuel intake. It's not as if lacking all this makes plants insensitive and unresponsive to their environment: they can control the direction of their growth, and can navigate toward what's good and away from what's bad like us, albeit very slowly and within a highly restricted area.

Before we descend into a meaningless "who's better—plants or animals?" argument, we should note the obvious point that our locomotory way of life must provide sufficient reward to outweigh the extra energy expenditure; so, too, must the sedentary lifestyle of plants pay off in spite of their thriftiness. Motility and nonmotility are just two different ways of being, and there's room on this planet for both. Which strategy ends up being favored by natural selection is entirely dependent, as always, on the given circumstances and the biology of the organisms in question. As if to prove this point, the history of our kingdom is peppered with instances of locomotory abandonment and reacquisitions. Our original point of distinction should therefore be reframed: what makes plants different from animals isn't their nonmotility per se, but their resolute dedication to a static existence. This is in spite of the fact that, surprising as it may seem, their single-celled ancestors were as fully motile as ours. Plants' locomotory reluctance is doubly curious, given their tremendously successful occupation of the terrestrial environment—the one place you won't find nonmotile animals, with good reason (more on this later). Something about plants caused the evolutionary door back to locomotion to slam shut. Or rather, it very nearly slammed shut. We're going to find that plants were never able to do away with locomotion entirely. In fact, locomotory considerations have dominated the evolution of the plant kingdom just as much as they've driven the evolution of us animals, just in a very different way. We'll be returning to these points later as well. First, however, we need to look closer to home, at the various options available to animals that feel like taking it easy.

## DISCOUNT TRAVEL

If I were to decide that locomotion really wasn't for me anymore, I'd die of thirst in about three days (five at a push), assuming some kindly passerby didn't bring me a drink. Giving up on movement isn't easy if what you need is elsewhere. But it's not impossible. Ours is a dynamic planet, and naturally occurring currents of air and water abound. With appropriate adaptations, coupled with being in the right place at the right time, these fluid conveyor belts can be used as free public transport. We've already encountered this strategy: rising air can keep soaring birds, such as vultures and frigate birds, aloft for hours or days, in some cases allowing them to travel hundreds of miles with barely a flap of their wings. Ocean currents can be used in a similar way. A number of shrimplike crustaceans called amphipods are known to feed on algae growing on the underside of Arctic sea ice: a productive if potentially unstable habitat (ever more so these days) that's liable to drift south and disappear come the summer, leaving the amphipods all at sea. The recent discovery of a batch of these creatures in deepwater plankton samples, collected near Svalbard, Norway, in the depths of winter by the intrepid Jørgen Berge of Svalbard's University Centre and his team, suggests how they might deal with this problem. By releasing their grip on the melting ice and sinking, the amphipods may be making use of a deep current that flows back toward the heart of the Arctic Ocean. Once there, they'd need only rise back to the surface in time for spring to find the lush (if frigid) ice-bound meadows once more. At least one of the two vertical migrations must involve some active locomotion, but compared to the relentless slog that would be needed to swim all the way back to the Arctic, the energy expenditure is trifling.

Fluid currents aren't the only way of getting a free ride. While it might sound risky, larger animals on the move can provide an effective taxi service for any creature tenacious enough to hang on. Many wingless terrestrial arthropods hitchhike in this way. Mites are especially famous for it, although their arachnid relatives the pseudoscorpions are more conspicuous. As their name suggests, these look like tiny scorpions, but without the stinging tail. Their pincers come in very handy when dangling from a fly or beetle in midair. However, the most remarkable freeloaders are marine. The remoras—a small group of medium-size fish—use a modified dorsal fin to attach themselves to sharks, whales, turtles, manta rays, or anything else that's large enough and moving, including human divers and ships. Its alleged attachment to one particular ship has granted the fish legendary status. Pliny the Elder, in his *Natural History*, blamed a remora for Mark Antony's fateful defeat at the Battle of Actium by Octavian (the future Emperor Augustus) in 31 B.C. The fish apparently added so much drag to Antony's flagship that the vessel was held securely in place, forcing its commander to abandon ship and flee. According to Pliny:

> Winds may blow and storms may rage, and yet [the remora] controls their fury, restrains their mighty force, and bids ships stand still in their career; a result which no cables, no anchors . . . could ever have produced!

It must be the most ridiculous excuse I've ever heard. Nevertheless, the fish ended up being named after the superpowers ascribed by Pliny: *remora* comes from the Latin, meaning "to hold back."

Of course, the remora would be a pretty useless hitchhiker if it caused its host to grind to a halt. It really wants to minimize its drag as much as possible—the higher the drag, the bigger the risk of being dislodged—so it always aligns its body with its host's direction of travel. The attachment mechanism itself is an evolutionary marvel. The supporting spines of a remora's dorsal fin have been transformed beyond recognition into a series of flattened strips that are hinged along their front edges. When not in use, these lie flat, but each can be tilted into a near-vertical configuration like the slats of a Venetian blind. When the modified fin is pressed against the side of a fish, this tilting has the effect of prying the two surfaces apart, creating a partial vacuum thanks to a rubbery seal that runs around the fin's perimeter.

Unlike arthropod hitchhikers, remoras aren't trying to get anywhere in particular. Rather, their objective is simply to keep pace with their host as cheaply as possible. Their ancestors probably behaved like today's pilot fish, which follow sharks, rays, and turtles to feed off the larger animals' parasites as well as any scraps left over from their meals or, failing that, their feces. Latching onto the creature just makes this job easier. Therefore, remoras are not usually regarded as parasites, although the extra drag is no doubt unwelcome. However, they're sneakier than they look: if the host is swimming fast enough, remoras open their mouth slightly to let the water current flow over their gills, thereby getting a bit of free ventilation as well as a free travel ticket.

Before moving on from hitchhikers, it's worth pointing out that the use of another organism's propulsive effort is not always so one-sided. The V-formation flight of geese, for example, arises because a flying bird generates an updraft of air behind and to the sides of its wings; any bird sitting in this updraft won't have to generate quite so much lift to support its weight. The only loser is the bird at the head of the V, but given that there is a regular change of leadership, all the birds end up benefiting. Bicycle pelotons work in much the same way, although in this case the benefit is the reduction in drag experienced by one rider when sitting in the wake of another. Admittedly, that is a somewhat artificial example, but the system has its natural equivalents in the shoals of certain fish and, remarkably, aggregations of swimming sperm cells. The most extraordinary example of a locomotory symbiosis, however, has to be that displayed by the single-celled protozoan *Mixotricha paradoxa*, which lives in the guts of termites. This creature appears to be covered in wriggling cilia, but these are actually symbiotic spirochete bacteria stuck to its cell membrane.

## STAYING PUT

All the locomotory-effort-saving techniques we've looked at so far presuppose that an animal needs to get from place to place, but that's not necessarily so. If there's a dependable supply of oxygen, food, and water in a single location, why bother? Many animals, unable to come up with a good answer to this question, have embraced a sedentary existence instead, by becoming ambush predators or fine dining enthusiasts that get by on minute particles sifted from the water or gleaned from the substrate.

It's rare for ambush predators to give up on locomotion—the ability to give a quick lunge at the last moment is usually too useful to discard—so they tend to look broadly similar to their more active cousins. Those ambushers that *are* rooted to the spot will have evolved from sessile ancestors.* It's the full adoption of the sedentary life by particle feeders that's had the most striking impact on animal form. Indeed, some of the creatures who've taken this path are barely recognizable as animals, which in hindsight is hardly surprising, given the dominant influence of locomotion on animal morphology. It's a road well traveled: almost every major animal group has sedentary representatives, and some, such as the sponges, are wholly so. All are aquatic, air being too rarified to support the steady rain of organic matter that fine diners need. Among the annelids we have tube worms, the mollusk flag is waved by bivalves (clams, mussels, and whatnot), and the cnidarians have their polyps. Even the locomotory titans the arthropods and chordates have a few workshy representatives, notably barnacles† and sea squirts.

With so many different starting points, there's a high degree of morphological diversity among the various sedentary groups. But there are also

* Among the entirely sedentary ambush predators are certain sea squirts that have turned one of their siphons into a prey-capture device.

† Barnacles are so un-arthropod-like that everyone thought they were mollusks until the 1830s, when they were found to have diagnostic crustacean larvae.

some common themes. Complex sense organs are always lost: there's no point painting a hyperaccurate picture of the environment if the most you can do is duck out of harm's way. Given that the brain's main job is to gather and process input from the sense organs for the purposes of navigation, this expensive organ tends to go as well. Sometimes the loss happens within an organism's life span. A tadpole larva of a sea squirt, if you remember from Chapter 4, has a rudimentary central nerve cord to control its swimming, but after diving headfirst to the seabed and gluing itself to the spot, it never swims again. Its central nervous system, now a pointless extravagance, is summarily digested.

As noted in Chapter 6, overt bilateral symmetry is another common casualty of a sedentary existence. When feeding on what's essentially nutritious dust, it often makes sense to spread the collecting gear over as wide an area as possible. Radial symmetry is the usual answer, so near-circular fan-like arrays of feeding tentacles are commonplace: corals, barnacles, hydroids, and many tube worms all use them. The situation is a little different for sea squirts and bivalves: these have internal food-collecting filters and generate a water current to bring in their particulate meals through a tubular siphon, like a vacuum cleaner.

However radical the departures from the ancestral motile form, there's not a single sessile species in whom the ghost of locomotion past isn't evident. For a start, many can still move around, even if their excursions extend no further than the confines of a tube. Even those animals that are firmly rooted to the spot can use their muscles to deploy or retract delicate tentacles or cower away from trouble. Beating cilia, originally locomotory structures just like muscles, often play a vital role in food collection, and not just by driving a current of water. Bivalves, like many fine diners, secrete mucus to help catch food particles, and the nosh-laden snot is literally reeled into the gut by a spinning cylindrical spool (a crystalline rod of protein) whose motion is driven by cilia.‡ Ironically, the repeated abandonments of locomotion in the animal kingdom wouldn't have been possible without the apparatus that was originally responsible for propulsion.

The hand of locomotion actually extends much further than this. I must confess that I've been taking an overly adult-centric view so far. Sitting in one place is all very well if the only thing you're worried about is stuffing your face, but sooner or later you have to square up to your reproductive responsibilities.

‡Fans of zoological trivia may like to note that this is a rare example of a natural axle.

It generally won't do to just drop your offspring next door. Space is at a premium for sedentary animals, so the youngsters need to get out of their parent's hair (or tentacles) and find a place of their own. Looked at another way, as the dispersal capabilities of one's offspring increase, so, too, will the numbers that survive into adulthood to reproduce themselves: in other words, good dispersal means higher Darwinian fitness. Therefore, it is effectively impossible for an animal to completely give up on locomotion. If the adults can't move, the onus of dispersal falls on the shoulders of the larvae, which are locomotion-capable in all the sessile groups we've seen so far. And yes, they do need to actively propel themselves rather than just drift around. They might be able to ride the currents for a time, but they'll need engines and sensors to reach a suitable settlement site, be they ciliated, like the larvae of *Platynereis* (and many other species), or legged, like the one-eyed larvae of crustaceans, which use the breaststroke to gain a measure of control over their destiny.

Of course, before a larva can even think about undertaking its voyage of discovery, a sperm cell must have found its way to an egg cell, unless we're talking about one of the rare wholly asexual animal groups. Sexual reproduction is the great locomotory leveler, although the dispersal powers of sperm are pretty limited. For a sedentary species, this usually means that an entire population must release its sperm and eggs at precisely the same time. These mass spawning events—often synchronized to take place within minutes despite only occurring once each year—require astonishingly accurate long-term time-keeping skills, and the exact combination of internal and environmental cues used are in many cases still unknown. Witnessing one of these orgiastic displays is therefore a matter of chance as well as judgment. I gather the experience can be quite magical—like being in the middle of a vast underwater snowstorm. By the way, if you assumed that direct copulation is a no-no for sedentary animals, you'd be wrong. Believe it or not, most barnacles do just this, and are thus endowed with the longest penis (in proportional terms) in the animal kingdom: up to eight times the length of the body.

## SECOND THOUGHTS

The sedentary strategy has worked out well for many animals, but it seems that, for some, the quiet life wasn't all it was cracked up to be. It certainly has its drawbacks: if local environmental conditions deteriorate, there's not much a sessile creature can do except reproduce if it can and hope for the best.

There's also the perennial squabble over living space, and the aforementioned sexual difficulties. Any of these factors could turn the tide of natural selection back toward motility. Fortunately, the many traces of an energetic evolutionary past inherited by the sedentary have allowed some of them to reacquire propulsive skills. The easiest method is to become reproductively mature while still a larva, skipping the sedentary adult phase entirely. Precociously mature larvae turn up in plankton samples every now and again, so this is a solid possibility. Indeed, as we saw in Chapter 6, Urbi—the founder of the great bilaterian radiation—may have evolved from a planula larva of an ostensibly sessile cnidarian-grade ancestor.

Larvae aren't the only way back to locomotion: there are a number of examples of sedentary adults that have become refitted for propulsion. However, the transformations wrought on their sessile ancestors' bodies have in some cases been so extreme that they now rank among the strangest denizens of our planet. We met the salps—jet-propelled sea squirts—in Chapter 4. Scallops are the bivalve answer to life in the fast lane. Their shell-closing muscles have been pressed into service as pumps that repeatedly squeeze jets of water to either side of the shell hinge when in panic mode. The resulting swimming style doesn't look pretty, but it serves well enough to keep them from the clutches of marauding starfish (the nemesis of many a bivalve).

Starfish themselves, along with their relatives in the echinoderm group, are the perfect study in postsessile locomotion. It's impossible to make any sense of them without bearing their tortuous history in mind, for when it comes to propulsion, they break every rule in the book. They're radially symmetrical (specifically, pentaradially—with five-rayed symmetry), and have neither a brain nor eyes. There's no consistent front end—any of the starfish's limbs can take the lead as the situation demands. In most species, movement is powered by a collection of hydraulic tube feet that hook up to a network of seawater-filled canals called the water vascular system, seen only in echinoderms. Horror of horrors, these are not metachronally coordinated. And if you thought starfish were weird, wait until you see the rest of the family. Brittle stars usually row along the seabed using two of their five limbs, but again, there's no consistent front end. Many sea urchins walk on their spines: their five limbs have effectively curled up and become incorporated into the body wall. Sea cucumbers at least look like bilaterians: they're rather like elongated sea urchins that have fallen over. However, they're no better endowed in the brain department than the rest of the group.

8-1: The starfish *Nymphaester arenatus* (top), crawling on its hydraulic tube feet, an extinct stalked crinoid (lower left), and a group of extant free-living featherstars in the species *Florometra serratissima* (right).

There's only one echinoderm lineage for whom radial symmetry, brainlessness, and the one-of-a-kind water vascular system make clear adaptive sense: the crinoids. The name comes from the Greek for "lily," which sums up their original form nicely: they were once all stalked, sessile creatures (as their excellent fossil record makes plain), with a crown of slender arms for sifting food particles from the water. You might meet a few species today that still look like this, but most modern crinoids have lost the stalk. Known as featherstars, they use rowing motions of the arms to either crawl or swim, despite the fact that their appendages are nominally feeding organs. None can keep it up for long, but the technique serves well enough to, say, find a safe, secluded spot in which to hide during the day. The tube feet, by the way, help out with food gathering, by flicking passing particles into cilia-lined grooves that run down each arm.

The modest locomotory skills of featherstars give us a clue as to what the other echinoderms are all about, for it's widely believed that the last common ancestor of the entire group was basically a sessile, filter-feeding crinoid. Bilateral symmetry, a brain, and complex senses were of little use to such a creature, and so all were lost. But, just as in featherstars, natural selection pushed some of the descendants of this sedentary ancestor back toward locomotion. The process may have started when an enterprising individual literally turned away from filter feeding and bent over on its stalk to collect food from the surrounding substrate. In this position, the tube feet could help the arms shuffle around. Once dedicated to this way of life, the stalk was nothing but a useless shackle, and was eventually erased by natural selection. The result was a starfish in the making.

Echinoderms stand as testimony to the convoluted routes by which natural selection may bring about a reengagement with locomotion. But to see the true apotheosis of the once-sessile, now-motile evolutionary pathway you have to look to the cnidarians, specifically the hydrozoan siphonophores. These are often mistaken for jellyfish (the Portuguese man-of-war is the group's most famous representative), but they aren't in fact single organisms at all. Each is a colony of polyps, albeit with a striking level of job specialization. Some individuals deal only with feeding, some are dedicated to buoyancy like the man-of-war's carbon monoxide–filled float, some to reproduction, and others are living weapons that take the form of long, trailing tentacles, packed full of microscopic stinging harpoons.

Siphonophores' locomotory skills go far beyond what would be expected of a colony. The Portuguese man-of-war has the simplest approach—it's blown around by the wind—although it's a surprisingly accomplished sailor: the balance of aerodynamic forces above water and hydrodynamic drag below keeps the sail-like float at an angle of attack of about 40 degrees. It thus maintains a consistent heading on a port or starboard tack, depending on

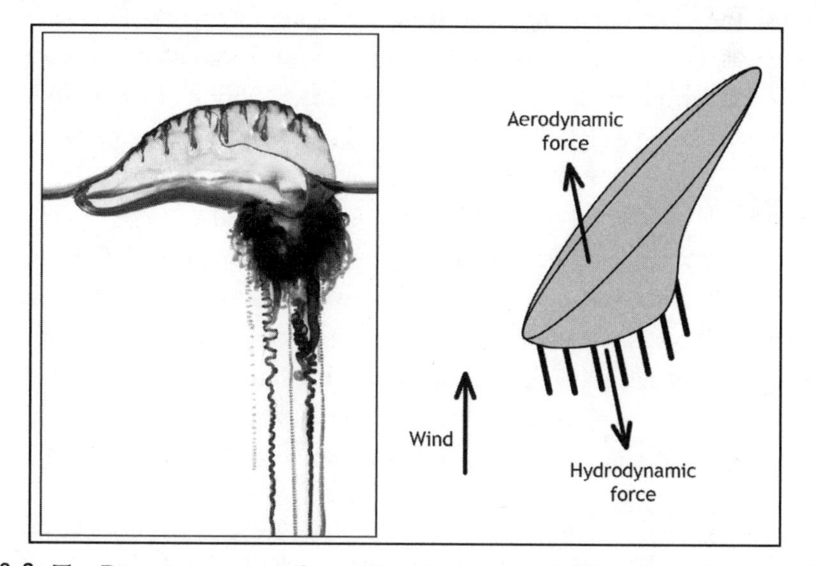


8-2: The Portuguese man-of-war *Physalia utriculus* (left), and the balance of forces on a sailing colony (right).

whether the colony is right- or left-"handed," and extracts a fair amount of lift from the wind (because the sail is vertical, the lift force is horizontal). The upshot is that the man-of-war can move surprisingly quickly. It's never in a hurry to get anywhere, but the extra speed helps the tentacles spread out behind, maximizing prey capture efficiency.

Despite the man-of-war's fame, its passive habits are unusual for siphonophores. Most are active swimmers, and would never dream of coming to the surface (although many have floats, these are of modest size and are used to regulate buoyancy, not bob around on the surface). Their strategy is an ingenious take on the multistage hydrozoan life cycle. Like standard hydrozoan polyps, active siphonophore colonies make swimming medusae, but hang on to their jellyfish children, pressing them into service as jet engines. The attached medusae are so stripped back that they can neither feed for themselves nor reproduce, but exist solely to propel the colony. It will by now come as no surprise to learn that the contractions of the medusa team are metachronally coordinated for maximum efficiency,* but in an emergency (if, for instance, one end of the animal is violently stimulated), everyone contracts together to accelerate the colony away from danger as quickly as possible. Some siphonophores can even do a reverse escape, which involves a collective swiveling of the medusa team's nozzles by 90 degrees before squeezing. For a colony to exhibit such a diverse locomotory repertoire requires an extraordinary level of coordination, underpinned by what are to all intents and purposes colony-wide neurons. The individuals are integrated so thoroughly that it's difficult to view a siphonophore as anything other than a single superorganism. They are surely the most wondrous evolutionary take on the reacquisition of locomotion.

## HIGH AND DRY

It's time we returned to our opening question. Now that we've got to know the sessile animals, we should realize that while plants appear to be immune to the usual locomotory imperative, this is an impression born of adult-centric chauvinism. Dispersal is as pressing for them as it is for us, and finding a way to deal with this issue on land despite having sessile adults has been the dominant concern of the plant kingdom ever since our photosynthetic cousins

---

* The Mexican waves of contraction work in siphonophores thanks to the colony's organizational scheme. Although individual medusae use a more or less typical cnidarian symmetry system, the team is strung out along the main axis of the colony in a somewhat bilaterian way—a wonderfully roundabout example of convergent evolution.

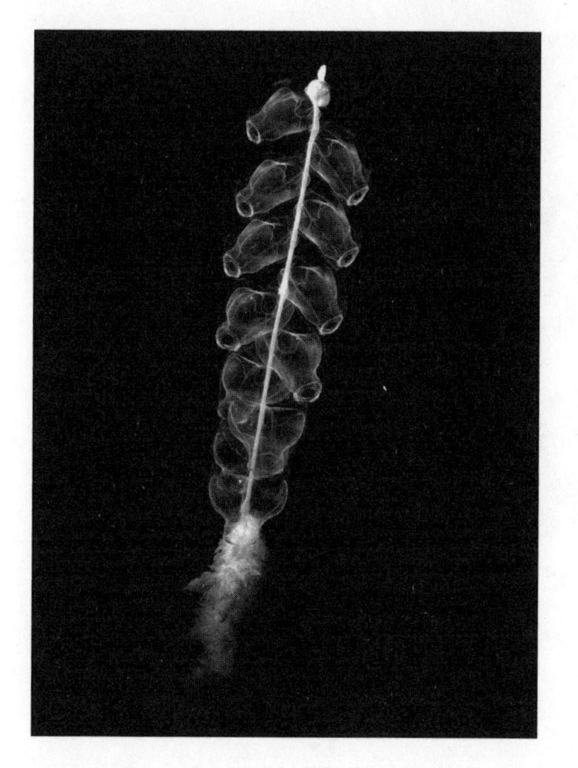

8-3: The siphonophore *Marrus orthocanna*.

first made landfall. Prior to the terrestrial invasion, plants had been using the standard aquatic sessile strategy for a while—settling down as adults and letting their sperm or embryos deal with dispersal, as seaweeds and primitive freshwater plants do to this day. Although this might seem totally ineffective on land, we might envisage that these tactics would work up to a point. A bit of rain could provide the necessary support that a sperm cell would need if an egg was close enough. But then what? Neither before nor after fertilization will the embryo be able to move from its birthplace. At best, we'd end up with generations of plants growing on top of their parents.

In primitive land plants, such as mosses, liverworts, and ferns, this generation-stacking is in a sense exactly what happens. Surprising as it may seem, all produce motile sperm that have to find their own way to egg cells (held in special chambers called archegonia) and the resulting embryos do grow on top of their parents. But the offspring look nothing like Mom and Dad. Plants, like advanced cnidarians, exhibit a multiphase life cycle. Sperm and egg cells are made by the so-called gametophyte (meaning "gamete plant," *gamete* being the catch-all term for sperm and eggs). In mosses, this is the conspicuous green cushionlike growth. The next generation, which grow from fertilized eggs, are the spindly, lamppost-shaped structures that

8-4: Two generations of a moss: the low-lying photosynthetic gametophytes and the tall parasitic sporophytes.

spring up all over the miniature moss meadow. These produce neither eggs nor sperm, but spores, and so are known as sporophytes. Spores are specially designed for long-distance dispersal. They're tiny enough for the gentlest of air currents to take them aloft, and the spore wall—invested with an ultra-resistant substance called sporopollenin—is often intricately sculpted, which increases drag (a good thing for a windblown disperser) while adding negligible mass. Should a spore land in a favorable environment (and they're produced in sufficient numbers to beat the appalling odds against this happening), it grows into a gametophyte that will in time produce sperm or eggs (or both) and begin the cycle again.

Spores, then, are the plant answer to motile larvae, but the responsibility for dispersal doesn't rest with them alone. In an aquatic environment, it may be enough for a parent to merely let go of a larva: local water currents and the creature's own engines will do the rest. On land, however, spores need more of a helping hand getting into the dispersive air currents. At the very least, they need to be released from a reasonable height, which ensures that they begin their journey away from the still air at ground level, while also increasing the length of time it takes to fall to the ground in the absence of an updraft. Here the onus is clearly on the sporophyte, which accordingly towers over its parent gametophyte, at least in relative terms. But growing tall isn't the only thing it can do. Many primitive plants have special mechanisms to forcibly eject the spores. Liverworts use elongate cells called elaters, which are scattered liberally through the spore capsule and twist violently when they dry out, flinging the spores to the wind. Similar catapults are seen in some ferns and their relatives the horsetails,* whereas most mosses limit themselves to merely opening the spore capsule when conditions are dry. The bog-dwelling *Sphagnum* mosses, however, have what's arguably the most explosive spore release mechanism of any plant. The walls of their lidded capsules are "designed" to collapse as they dry, which increases the air pressure within until it reaches a level roughly equal to what you'd measure in the tires of heavy trucks. The lid is then blown off with an audible click, and the spores are launched upward by about 4 inches.

Unlike plants, fungi weren't endowed with a handy multistage life cycle prior to their colonization of land; but because they faced the same difficulties of terrestrial life without locomotion, they essentially had to invent their own version from scratch. Their strategy involved first doing away with the troublesome water-loving sperm. The responsibility for finding a mate now rests with the network of threadlike hyphae that constitute the bulk of a fungus (a mushroom is merely the tip of the iceberg). Sexual union occurs when the hyphae of two different networks make contact and fuse—a mushroom or bracket or whatever is the chosen type of fruiting body then grows from the united hyphae. This structure is the fungus equivalent of a plant's sporophyte generation, whose shed spores grow into a new hyphal network if they're lucky enough to land in a suitable location.

As in the primitive plants, fungi have come up with a number of ingenious spore-release mechanisms to disperse their windblown children as far and wide as possible, from the ballistic spore guns of elf cups to the raindrop-powered spurt system of puffballs and earth stars. In both fungi and plants, however, there's simply no substitute for sheer height, hence the appearance of the gigantic *Prototaxites* fungus in the early Devonian that we met in Chapter 3. Plants stood to reap the same benefits, as long as they ensured that their sperm could still swim to their eggs. With their multistage life cycle, this was easy enough: the sporophyte could get as tall as the strength of its stems (and anchorage of its roots) permitted, as long as the sexual gametophytes kept low and moist. In ferns, this is exactly what we see. The leafy fronds constitute the sporophyte, while the gametophyte from which it grows is so inconspicuous that it's rarely noticed.

---

* Horsetails have an extra trick: their spores each carry four elaters, whose humidity-responsive twisting causes them to walk and even jump.

The ferns' height-based dispersal strategy, while successful, is somewhat antisocial: a vertically challenged plant in the shadow of its taller neighbor will have the majority of its light stolen, which is a big deal if you're photosynthetic. Relocation being ruled out, there are two solutions: the shaded plant either has to become a more efficient photon harvester or it must try to beat its neighbor at its own game by growing even taller. Thus was the stage set for a relentless botanical arms race that would change the very face of the Earth. The early phase was a relatively polite affair, because the first land plants were structurally equipped for only modest vertical growth. All that changed when, toward the end of the Devonian period, plants discovered that lignin—an ancient antifungal biopolymer that they'd evolved long before they colonized the land—had extraordinary strength when laid down in sufficient quantities. We call such excessively lignified plant tissue *wood*, and once this was on the scene trees started shooting up all over the planet.

This was when the light war got nasty. Normally, most of an organism's carbon is returned to the atmosphere when it decays, but at the height of the conflict, nothing existed that could break lignin down, so it all got buried instead. For millions of years there was a one-way flow of carbon from the air into the ground, where it turned into coal. Within 20 million years, trees had stripped the atmosphere of 80 percent of its carbon dioxide, chilling the Earth sufficiently to initiate a major ice age. But the damage didn't stop there. The extensive roots that supported the grotesquely distended aboveground growth caused an unprecedented degree of weathering, leading to nutrient enrichment of waters on a global scale and widespread algal blooms. The collective decay of the algae caused a precipitous drop in atmospheric oxygen concentrations, down to a low of 12 percent if certain atmospheric/geological models are correct. This crisis was later reversed as terrestrial plants pumped out the oxygen waste of photosynthesis—indeed, by the end of the Carboniferous period, levels had risen to an all-time high of about 35 percent—but the damage was done. By the end of the Devonian, a full 75 percent of living species had been wiped out, all because plants couldn't be bothered to just get off their roots and move. Having said that, things could have been a lot worse. Toward the end of the Carboniferous, the mushroom-forming group of fungi finally worked out how to degrade lignin, putting a halt to the inexorable burial of carbon dioxide. In so doing, they may have saved the planet from freezing over entirely—something to consider next time you bite into a mushroom!

## AEROPLANTS

With their potentially soaring heights and minuscule spores it might be supposed that the plants' dispersal woes were over. But their original reproductive technique suffered from one major drawback. If a sperm has to swim to an egg outside the confines of the body, an organism can never be wholly terrestrial. No matter how far the spores are disseminated, plants that adhere to such a strategy can only ever occupy habitats at the wetter end of the spectrum, where you'll find mosses and ferns to this day. To strike out further, something had to be done about this. And indeed it was, as is immediately apparent from the current geographical reach of plants, which extends to some of the driest places on Earth. Their solution was to turn the waterborne sperm into airborne spores, although not directly—that would have been too drastic a transition to pull off. Instead, evolutionary attention was focused on the gametophyte stage, which bridges the gap between sperm and spore. The first step was a division of labor in the gametophyte population, whereby sperm were made only by one kind, eggs by the other. In other words, there was a transition from a bisexual gametophyte (which is used by most ferns) to separate males and females. Once that was done, the male gametophyte could be stripped right back until it was little more than the spore from whence it came. In effect, it became an airborne sperm, although we'd know it as a pollen grain.

Once sperm had become air-worthy, there was little need for female gametophytes to grow as separate ground-hugging plants (although there are a few oddballs that still follow this scheme). Instead, each could undergo a modest amount of development up on the sporophyte—enough to make an egg cell and some nutritive tissue to support the growth of the future embryo—and wait for the pollen to come to it. Thus were seeds born, whose arrival neatly got rid of the need for swimming sperm, and also ensured that sporophyte embryos always got off to a good start in life. Fossil seeds tell us that this reproductive/locomotory strategy evolved toward the end of the Devonian period (before the main phase of the light war), and seed plants have never looked back—they now dominate the land, from tundra to rain forests.

Of course, no strategy is perfect. For all their advantages, the first seed plants were faced with a dilemma: laying down lots of food reserves for the embryo is great, but the increased mass of the seed must surely impact on dispersal. As the saying goes, the bigger they are, the harder (and more important, the sooner) they fall. The earliest seeds were therefore small enough that the wind could carry them. Plenty of seed plants today, such as orchids,* still start out tiny for the same reason. But you don't have to be minuscule to extend your stay in the air. Remember, the reason large objects have such a hard job staying aloft is nothing more than the mismatched scaling of drag, which depends on an organism's area, and weight, which depends on its volume. The simple growth of a bit of fluff to increase drag without significantly increasing weight can do a fantastic job of keeping smallish seeds airborne in a light breeze. Dandelions are famous users of this strategy.

Larger seeds need something more than fluffy parachutes. Samaras—the winged seeds (technically fruit) of maples and the like—use lift as well as drag to extend their dispersal range. In a way, they're even more adaptively impressive than the flying animals because, lacking brains or muscles, they're "designed" in such a way that they automatically find the optimal configurations and motions for efficient flight. Yet they're deceptively simple things: a maple samara is little more than a flat blade with a weight (the seed) at one end. You'd surely expect that if dropped from a great height, it would plummet to the ground seed-first; instead, it starts spinning around like a miniature helicopter. The key to this strange behavior is the way the samara's mass is distributed. It's concentrated not just at one end, but also toward the leading edge, thanks to the densely packed veins that run along the front of the blade. As a result, when the samara starts to fall, it tumbles forward about its long axis as well as tilting seed-down. These rotations interact as in a gyroscope, initiating the spinning motion, at which point centrifugal forces acting on the seed and blade reverse the seed-down tilt. The samara is now effectively gliding in a tight spiral, and the lift it generates can slow its descent by 80 percent. That alone would do it no good, but in the presence of a decent side wind (and the samara's attachment to its parent tree ensures its release only when it's windy), a slow descent could mean the difference between life away from its parent and death by shade, nutrient starvation, or insect jaws underneath it (a mature tree houses a fearsome community of microherbivores just waiting to pounce on any juicy youngsters within striking distance).

Samaras have evolved many times in seed plants. Some, such as the pine and hornbeam versions, look and function much like maple fruit. Others use different aerodynamic techniques: the tree of heaven samara, for instance, has its seed slap bang in the middle of a propeller-like blade, so

* Orchid seeds are so small that they need the help of fungal symbionts to germinate.

8-5: Three samaras with different aerodynamic strategies: from a maple (left), an ash (center), and a tree of heaven (right).

8-6: The gliding seed of the Javan cucumber *Alsomitra macrocarpa*.

spins backward on its axis as its falls,† generating lift like a spinning tennis or golf ball. The champion botanical aviators, however, are the diaphanous, wing-shaped flying seeds of the Javan cucumber. These beautiful, 6-inch-wide objects, released from opened-bottomed husks, are true gliders, and good ones at that: the typical glide angle is a respectable 12 degrees. Each maintains a stable configuration in flight thanks to the forward placement of the flattened seed and the relatively flexible trailing edge of the wing, which

†It spins backward because its aerodynamic center, as is usual for airfoils, lies closer to the leading edge than to the trailing edge—ahead of its center of mass.

deflects upward when aerodynamically loaded and acts like the tail of a man-made aircraft.* Its ultralightweight construction means it can't deal with strong winds, but then, it doesn't have to—the plant is a climbing liana of the Southeast Asian rain forests (the great glider nursery) and its seeds do much of their gliding beneath the canopy.

These various aerodynamic tricks impress us with their elegant sophistication, but there are plenty of other ways for a plant to disseminate its seeds. Ballistic ejection works well for the small. Water can also serve if available, the only proviso being that the seeds must remain at the surface while in transit (unless we're considering a submerged plant, such as a water lily). Although surface tension can be exploited by seeds that are sufficiently small and waterproof, flotation using air spaces or oils is the usual adaptive solution, which allows some very large seeds, such as coconuts, to travel thousands of miles away from their parents. Then there are the tumbleweeds—the catch-all term for a number of convergently evolved plants whose aboveground growth detaches and goes for a spin, spreading seeds as it goes (this obviously only works in open habitats, such as deserts). But the most common alternative to air travel is the taxi option. Terrestrial plants may have given up on locomotion, but their animal neighbors haven't, and with sufficient enticement or subterfuge they can be drafted in to disperse plants' seeds for them. It's the plant version of hitchhiking.

## LOCOMOTION BY PROXY

The simplest way for a plant to get its seed transported by an animal is to stick it to the creature's body, using glue or Velcro-like hooks.† More often, though, animals are bribed to carry seeds away with tasty, fleshy fruit. We vertebrates are the favored couriers, thanks to our large size—a butterfly won't get far with a cherry stone. The usual idea is for the seeds to be eaten along with the fruit, taken away in the animal's digestive system, and ejected some distance away, but it doesn't always go according to plan. Having a great mass of indigestible seeds rolling around inside one's gut can be quite an inconvenience, especially for birds, so some just strip the edible parts of the fruit away and drop the seeds where they find them. Mammals are less bothered by the

* The Javan cucumber samara inspired one of the world's first piloted gliders, built by Austrian aviator Igo Etrich in 1903. Its inherent stability made it an attractive model at a time when we were still feeling our way with aerodynamics.

† Hooked fruit—burrs—were in fact the inspiration for Velcro.

roughage, but often have the irritating tendency of dumping an entire batch of seeds in one place when they defecate, making a mockery of the whole competition-avoidance rationale for dispersal. Nevertheless, the fruit-as-bribe strategy clearly works well enough, especially in the windless understory of tropical rain forests, and it has a long and distinguished history: some fossil trees from the Carboniferous period appear to have had fleshy fruit. We primates have played a special part in the unfolding of this history. Our ancient ability to negotiate the fine branches upon which fruit are usually borne, our unusually good color vision (by mammalian standards) that lets us pick out ripe fruit from the background foliage, and our shunning of flight with all its weight-related worries made us unusually effective plant dispersal agents.

One of the hallmarks of true self-directed locomotion is the ability to choose one's destination, and for all the ingenuity of the various seed dispersal strategies, this is a skill that plants never acquired. Or did they? If control over a seed's destination was really beyond plants' capabilities, it's doubtful that mistletoes would have evolved. These are semiparasitic plants—they're photosynthetic but rely on a host plant for water and nutrients. That wouldn't be a big deal, if it weren't for the fact that mistletoes need to establish themselves way up in the host canopy. This gives them access to light, but how do their seeds get "planted" up on a branch? The answer lies in the sticky coating of their seeds, which remains intact even after passing through a bird's digestive system. When the time comes for the seed to be voided (either by defecation or regurgitation), its sticky jacket clings to whatever orifice it came out of, forcing the bird to wipe the seed onto a nearby branch. The young mistletoe is now perfectly placed to plug into its new host.

For the brief period of time between a mistletoe seed's consumption and its deposition, the bird that swallowed it effectively becomes an extension of the plant, or part of its extended phenotype, to use the term coined by Richard Dawkins to describe any influences of an organism's genome that stretch beyond the organism itself. In a way, the simple manipulative strategy of mistletoes grants their seeds animal-style propulsive abilities. The same could be said of any plant seeds that use animal couriers, although in most cases the relationship is more tenuous and the results less predictable. More often than not, vertebrates just seem to be rather difficult for plants to manipulate. The same cannot be said for insects. The relationship between some insect species and the plants they pollinate—usually mediated by flowers—can be so tightly intertwined that neither can now survive without the other. No system illustrates this mutual interdependence better than figs and their tiny wasp tenants. A fig "fruit" is actually an inside-out cluster of miniature flowers within which the wasps begin their lives. The males, which are usually wingless and live out their whole lives within the fig, hatch first, and fertilize the females while they're still inside their floral nurseries. Each fertilized female then exits her home fig and flies to another, entering via a hole so narrow that her wings are often torn off. Unfazed by this mutilation, she quickly gets to work laying eggs in some (but not all) of the flowers, pollinating as she goes. This is her last act—the fig becomes her tomb. Any pollinated flowers that escaped the wasp's egg-laying now become seeds, which will later be dispersed by whatever eats the mature fig, usually a bird. A single species of fig wasp (and there are many) is usually dedicated to a single species of fig, so to all intents and purposes the wasps are the plant's eyes, legs, and wings, delivering its sperm exactly where it needs to go. Would that the same could be said for the birds: one study found that only 6.3 percent of a fig tree's crop of seeds are dispersed undamaged by their unreliable couriers.

The employment of arthropods by plants to help them have sex might be thought to have arisen with flowers. These certainly make the pollinators' job much easier: the petals act as advertising billboards, with distinctive logos to help an insect find the right species. It's also in flowers that we find nectaries, which dish out the sugary fuel that their pollinators seek. Nevertheless, there's good evidence that the arthropod/plant association was in place long before the first petals unfurled. Triassic-age fossil cycads—an ancient lineage of seed plants related to conifers—have been found with suggestive pollen-laden arthropod feces inside their cones, and some cycads are insect-pollinated to this day. In fact, recent studies indicate that the interaction may date back even further—to the days of the very first land plants. Experiments carried out by Nils Cronberg of Lund University, Sweden, have revealed that lowly mosses make use of arthropod couriers to help with sperm transport. He placed a male and female moss gametophyte onto a bed of plaster of paris, separated by a gap of about an inch (the plaster's absorbent properties ensured that no swimmable film of water straddled the gulf). Cronberg found that sporophytes would only grow from the female gametophyte in the presence of mites and springtails, so these tiny animals must have carried sperm from male to female. The pollination parallels don't end there. Todd Rosenthiel and his colleagues at Portland State University found that female mosses emitted volatile chemicals that attracted nearby springtails, in the same way that flower perfumes attract winged insects. The egg-holding archegonia may even provide a sugary reward analogous to floral nectar.

## WALLED IN

I hope you can now appreciate that plants' inability to wander around has had a singularly dominant influence on their evolution. To cut it in a world where some kind of locomotion (broadly defined) is essential at some point in the life cycle, they've had to come up with all kinds of tricks, from the simple (e.g., growing tall) to the complex (e.g., aerodynamic fruit). From an extended phenotype point of view, whenever those strategies involve the employment of animals to provide their movement, plants could be said to have evolved full-blown, self-propelled locomotory powers. But they've done it via the most absurdly convoluted route. Why couldn't plants just take a leaf out of the siphonophore book and rediscover motility?

This would be a good point to revisit the Venus flytrap. Its murderous behavior is closer to animal-style movement than is anything else in the plant kingdom, putting it in a unique position to shed light on the limits of plant motion. It has long been known that the basis of the trap closure mechanism is a rapid change of curvature of the leaf lobes (the "jaws" of the trap). When open, each lobe is convex on the inside, snapping to concave when the trap is sprung. For an animal, such a change is easy to bring about. Indeed, the evolutionary success of our own chordate group largely rests on this kind of movement. We use longitudinal muscles on either side of the incompressible notochord or spinal column to drive the convex-to-concave bending. But plants, of course, have no muscle, so their cells can't actively pull. However, their cells can push. Between the upper and lower epidermal cell layers of the leaf lobes lie the internal mesophyll cells, which are under high internal pressure. While the trap is open, this pressure is restrained by tension in the superficial epidermal layers, just as the internal pressure of the notochord is restrained by its fibrous jacket. When the trap is activated, the cells of the lower epidermis are altered in such a way that their restraining action is relaxed, but only in the direction perpendicular to the trap hinge. The pressurized mesophyll is now free to stretch the lower epidermis in this direction. It doesn't extend by an enormous amount—only about 7 percent of its initial width—but that's enough to snap the lobe from convex to concave.

The Venus flytrap mechanism could never work in an animal: our cell membranes are too flimsy to resist the sorts of pressures involved. Try to pump up an animal cell and it ruptures (the cells in a notochord are exempt from this rule, thanks to the notochord's tough external jacket). Plants can cope with this sort of thing because their cells are surrounded by cellulose-reinforced walls, which can withstand hefty pressure differences between the

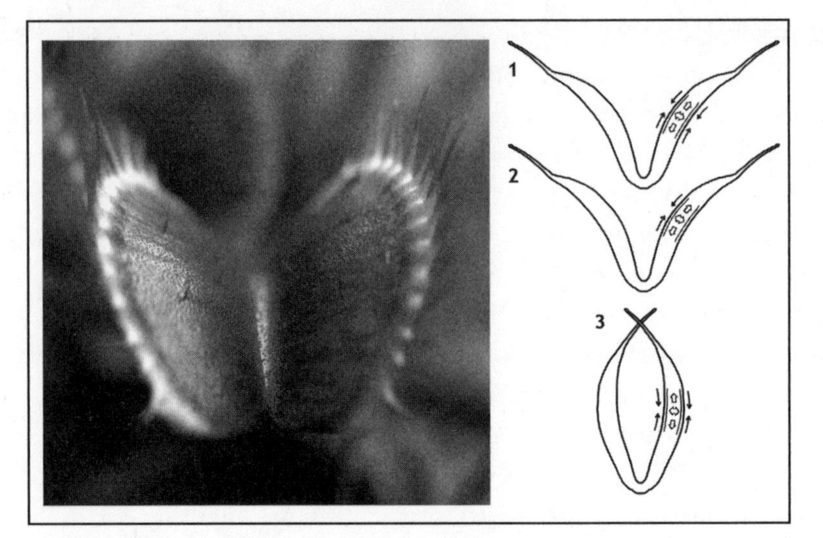


8-7: The closing mechanism of a Venus flytrap. An open, unstimulated trap is rather like a drawn bow, with pressure in the internal mesophyll cell layer resisted by tension in the outer epidermal layers (1). When a trigger hair is touched twice in quick succession, an electrical impulse propagates rapidly across the leaf, which loosens the lower epidermal layer (2). Like releasing a drawn bowstring, the now unrestrained mesophyll pressure causes the leaf to snap from convex to concave, and the trap slams shut (3).

interior and exterior. This so-called turgor pressure, caused by the absorption of water by a cell, is the basis of nonwoody plants' structural integrity: deprive them of water and they droop. But turgor pressure can also do mechanical work, for if certain chemical bonds in the wall of a turgid cell are broken, the pressure difference will cause it to expand. What's more, the precise orientation of the cellulose fibers in the wall dictates the vector of the expansion. This is how the epidermal cells of the Venus flytrap extend only in a direction perpendicular to the hinge, with the minor complication that the movement is largely powered by the expansion of the underlying mesophyll cells rather than the intrinsic turgor pressure of the epidermis itself.

In a sense, the forceful extension of a plant cell is merely the opposite of the forceful shortening of an animal cell that's caused by its internal muscle fibers (more on these in the next chapter). But there's an important difference. Although a pressure-powered plant cell can elongate by an enormous margin (over a thousandfold, in some cases), these large-scale changes always involve the synthesis of new wall material, and thus are irreversible. This situation is roughly analogous to that of a hypothetical animal muscle that lacks antagonists: such a muscle would be strictly one-use only. For a plant, reversible movements are vastly more limited in range, and are only mechanically practical if amplified by tissue geometry, as in the Venus flytrap.

Finally, we have an answer to our question. The reason that plants are so resolute in their rejection of locomotion is because long before they became big and terrestrial, natural selection built walls around their cells. We know that the wall has such an ancient provenance because the land plants' close single-celled aquatic algal relatives are similarly endowed. Interestingly, many of these algae are fully capable of self-propulsion—a cell wall is no impediment to locomotion as long as you're small enough to use cilia, which can poke through the barrier. Although it may seem odd to find photosynthetic organisms motoring around, such creatures are likely to find self-propulsion useful more often than not, simply because they need to stay close to the surface of the water to get enough light.*

When plants' single-celled algal ancestors became multicellular, the old ciliary movement strategy would have served them perfectly well at first. Indeed, going by today's simplest multicellular algae—the hollow, spherical members of the *Volvox* group—ciliary propulsion can be particularly effective for the multicelled, because multiple beating cilia can be operated metachronally in a way that one cilium obviously can't (motile algal cells rarely have more than two cilia). As a result, the locomotory powers of *Volvox* are a match for those of, say, sponge larvae (*Volvox* even uses the same spiraling phototaxis technique that we looked at in the last chapter). That, however, is as good as it can possibly get for a creature with cell walls. Axons cannot be cultivated when cell membranes are so straitjacketed, so the nervous takeover of ciliary propulsion that happened in the early days of the animal kingdom was not an option for the plant line, despite the fact that plant cells have the ion pumps and channels that might support electrical communication (as the Venus flytrap conclusively demonstrates). And, of course, the gradual animal-style switch to muscle power, via an intermediate muscular steering system, was impossible. Hence, as body size continued to increase in the ancestral plant line beyond the point at which ciliary propulsion became ineffective, the only option was to give up on propulsion entirely and delegate all locomotory responsibilities to the parts of the life cycle that were still small enough to fulfil those duties—sperm and spores. The story was much the

* Nonlocomotory single-celled photosynthesizers do exist, such as diatoms: they rely on turbulence (sometimes coupled with the manufacture of low-density oils) to reverse their inevitable tendency to sink to the bottom.

same for the fungi: like plants, their cells are walled, so their ancestors faced the same locomotory troubles when they made the leap to multicellularity. We animals, however, thanks to our wall-less cells, were presented with a unique opportunity to continue our locomotory career in a multicellular context. While a more expensive way of life than that enjoyed by typical multicellular life-forms, it also offered bigger rewards, as it enabled the directed pursuit of concentrated packets of food—other organisms.

When all's said and done, the vast morphological gulf between the locomotory animals and the (largely) static plants and fungi is thus ultimately attributable to a simple difference of opinion about whether a living thing should build a wall around itself or not. In the next chapter, we're going to find out what precipitated that minor disagreement (as usual, locomotion holds the key, though probably not how you'd think), and investigate the origin of the two great propulsive engines of multicellular life—cilia and muscles—over whose deployment the cell wall had such a major impact. It's time to enter the alien world of the single-celled.

9

# Exodus

## In which we see how it all began

> A journey of a thousand miles begins with a single step.
>
> —Lao-tzu, *Tao Te Ching*

As scientific revolutionaries go, Antoni van Leeuwenhoek cut a singularly modest figure. Born in 1632 to a Dutch basket weaver, he took up trade as a draper, opening his own shop in Delft around the time of his twenty-second birthday. Here he would stay for the rest of his ninety-year-long life, and for the first two decades of his career he seemed set to lead the life of an unremarkable if respectable businessman. However, during that time he developed an unusual hobby that would not only change the course of his life but profoundly alter our very conception of the living world and our place within it. By day he may have sold buttons and ribbons, but in his spare time, Leeuwenhoek became the preeminent microscopist of his generation.

We're not absolutely sure what drew Leeuwenhoek to microscopy, although many think that the catalyst was Robert Hooke's best-selling masterpiece *Micrographia*, published in 1665. Significantly, along with the famous, gigantic foldout illustrations of a louse and a flea, Hooke had depicted a section of cloth, which may have piqued Leeuwenhoek's interest, textiles being his stock-in-trade. Hooke's exquisite engravings showed the Dutchman that

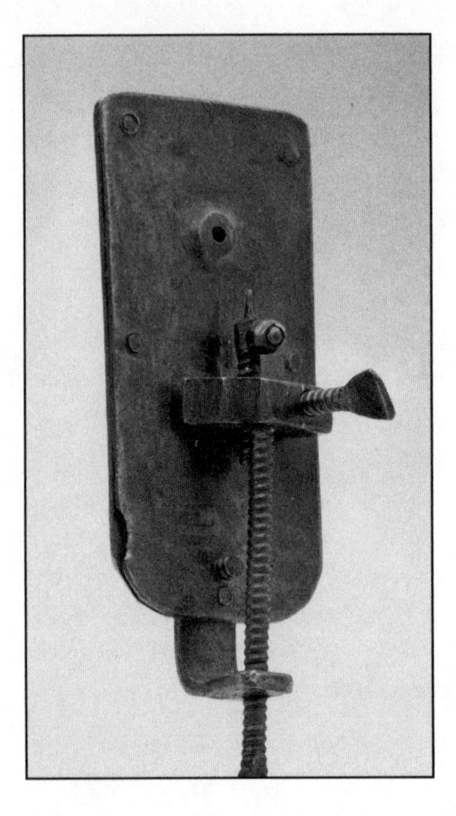

9-1: One of Leeuwenhoek's many microscopes. The specimen is held on the pin and observed through the tiny circular aperture.

a hidden world was waiting to be revealed: a world requiring no voyage of discovery to find—only the right equipment to bring it within the range of human perception. The temptation was too much to resist, and by the 1670s Leeuwenhoek was making his own microscopes. The intricate, multilensed devices used by Hooke being beyond his means, these were essentially customized, single-lens magnifying glasses, each consisting of two small brass plates that held the lens, and a pin on which to mount the specimen. The whole thing was less than 3 inches long, and was used by holding it up to the eye, with a window in the background. To achieve high magnifications, the lens had to be as tiny as possible (the smaller the radius of curvature, the greater the magnifying power)—this was where Leeuwenhoek demonstrated his true ingenuity. His technique (which he kept a closely guarded secret) involved heating a rod of glass over a flame and drawing it into a thread until it broke: the broken end was then melted into a tiny glass sphere. The best of Leeuwenhoek's scopes could magnify an object up to five hundred times (as we know from modern replicas)—a considerably superior performance to anything the contemporary compound microscopes could do. It would be a full 150 years before those more complicated devices caught up.

Armed with his optical equipment, Leeuwenhoek subjected all kinds of items to the closest scrutiny: fungal spores, a louse, and the sting of a bee were among the first. He communicated his findings by letter to Henry Oldenburg, secretary of the newly formed Royal Society in London, following an introduction by a mutual friend. This was in spite of the fact that England and Holland were technically at war at the time, thanks to England's capture of New Amsterdam (renamed New York) in 1664. Fortunately, the learned men of the Royal Society considered such conflicts to be subordinate to the aims of science, and saw no harm in the correspondence (although Oldenburg was often cautious enough to sign his letters with an anagrammatic pseudonym—Mr. Grubendol—to avoid attracting any unwelcome attention from the authorities).

One of Leeuwenhoek's early letters, written in 1674, details his observations of a cow's eye, along with a few comments on salt and the differences between Flemish and English earth. But it was the last paragraph that was the most significant. In it he writes about the contents of a glass vial of water that he'd collected from a lake near Delft:

> I found floating therein divers earthy particles, and some green streaks, spirally wound serpent-wise, and orderly arranged, after the manner of the copper or tin worms, which distillers use to cool their liquors as they distil over. The whole circumference of each of these streaks was about the thickness of a hair of one's head . . . and there were very many small green globules as well. Among these there were, besides, very many little animalcules [little animals], whereof some were roundish, while others, a bit bigger, consisted of an oval. On these last I saw two little legs near the head, and two little fins at the hindmost end of the body.

This was Leeuwenhoek's first step into the wonderful world of the microorganisms, never before seen by human eyes. Those spirally wound green streaks could only have been the unique helical chloroplasts of the filamentous alga *Spirogyra*. The "animalcules" with two little legs and two little fins may have been rotifers (wheel-bearers)—tiny animals named for their crown of beating cilia that give the illusion of rotary movement. The others, despite their roving disposition, weren't animals at all, but self-contained, single-celled life-forms.

From the moment *Spirogyra*'s twisted chloroplasts came into focus, Leeuwenhoek was hooked, and the following summer he embarked on a research project to find out as much as possible about these hitherto unseen creatures. He collected rainwater, river water, well water, ditch water, and seawater, observing a few drops from each sample repeatedly over a number of days to see how the miniature ecosystem changed. He also discovered quite by accident that an infusion of peppercorns became a veritable playground for microscopic life after a couple of weeks of soaking. His original plan had been to find out what caused the peppery flavor, but after seeing how his "animalcules" thrived on the decaying matter, he tried infusions of ginger, cloves, and nutmeg as well. He also correctly guessed, with great prescience, that dental plaque was a good place to look: he examined his own (he was rigorous in his dental hygiene, rubbing his teeth daily with salt) and also some scrapings from the "uncommon foul" teeth of an old man who had never washed his mouth in his life, but who drank brandy every morning. Leeuwenhoek wondered whether the tiny creatures that he'd found—undoubtedly bacteria—would be able to survive "such continual boozing" (needless to say, they could). On another occasion, he observed a sample of his own excrement during a bout of diarrhea, and by his description seems to have identified the pathogenic microbe *Giardia* therein. While he didn't at the time make the link between the organism and his symptoms, his work was an important step on the way to the acceptance of the germ theory of disease—one of the most important breakthroughs in the history of medicine.

In all his studies, two things impressed Leeuwenhoek more than any other. First, the microscopic organisms were often present in unimaginably vast numbers: even his most conservative assessments sometimes returned estimates of many thousands in a single drop of water. To our modern ears, such a figure probably doesn't sound that shocking, but in the seventeenth century this was revolutionary talk, and Leeuwenhoek felt compelled to let some local officials have a look through his microscopes to confirm that he wasn't making it all up. What Leeuwenhoek found most enthralling of all, however, was the sheer liveliness of his microcosms. In a letter dating from 1676, he observed of one of his pepper infusions that

> the whole water seemed to be alive with these multifarious animalcules. This was for me, among all the marvels that I have discovered in nature, the most marvelous of all; and I must say, for my part, that no more pleasant sight has ever yet come before my eye than these many thousands of living creatures, seen all alive in a little drop of water, moving among one another, each several creature having its own proper motion.*

* The entire letter, translated by Clifford Dobell, can be found in his book *Antony van Leeuwenhoek and His "Little Animals"* (1932).

For our locomotion-inclined purposes, it's this second observation that's the more significant. The world of the single-celled is truly restless. Of the dozens of major microbial groups, few indeed could be described as nonmotile, and even in those rare cases, there's usually a locomotory phase lurking somewhere in the life cycle. While the energetic nature of microscopic lifeforms was probably regarded by Leeuwenhoek as little more than a delightful quirk of God's creation, in the present-day light of evolution his discovery has added significance. He couldn't have known it at the time, but as Leeuwenhoek gazed through his microscope he was seeing the primeval underbelly of life on Earth. Before multicellular organisms evolved a little over 1.2 billion years ago, our planet belonged to these wriggling, roving animalcules. Indeed, it was their microscopic escapades that laid the groundwork for the macroscopic world that we know and love. That would be reason enough to get to know them, but on top of that, as the world's first movers they offer us the key to understanding the most important transition of all in the long history of locomotory evolution: the very origin of locomotion itself. We'll come to that momentous event in due course. First, though, we need to learn some home truths about how movement works at the scale of the single-celled, for it's not as easy as it looks.

## SWIMMING IN SYRUP

Throughout the book we've seen again and again how an organism's size dominates its locomotory options, thanks in most cases to the mismatch between the way an object's area and volume increase with size. Large fliers need relatively big wings: the lift depends on area (among other things), whereas the weight that the lift must balance depends on mass. A big terrestrial animal needs a relatively chunky skeleton for a similar reason: to keep the weight-induced stress within acceptable limits, the cross-sectional areas of the supporting bones must increase disproportionately. For an aquatic organism, however, it's quite easy to achieve near weightlessness, so one might conclude that area-to-volume ratios are of little relevance, and that size doesn't matter underwater.

If only it were that simple. In truth, size matters a great deal for swimmers, thanks to the mercurial nature of drag at different scales. When I first introduced this force back in Chapter 3, I came at it via Newton's second law of motion, imagining drag to be caused by innumerable collisions of fluid particles with the front of a moving object that weren't balanced by an equivalent set of collisions at the back. There's nothing wrong with this picture, but as a definition of drag it's incomplete. A moving object feels a retarding force not just because it's having to shoulder fluid out of the way—it also experiences friction as the fluid shears past its surface. It's as if there were two flavors of drag. The one we're familiar with depends on the inertia of the fluid barged aside, so is proportional to $\rho \times A \times v^2$, as we saw in Chapter 3. The frictional drag, on the other hand, depends on the surface area of the object and the viscosity (stickiness) of the fluid $\mu$ (Greek letter mu), as well as the steepness of the so-called velocity gradient near the object. This gradient exists because the relative fluid velocity at the surface is, to all intents and purposes, zero, rising to its nominal, free-stream value some distance away. Putting all that together, the frictional drag is proportional to $\mu \times A \times v/l$, where $v/l$ describes the velocity gradient in very simple terms. The ratio of these inertial and viscous forces is an enormously important quantity called the Reynolds number (Re), after fluid mechanicist Osborne Reynolds (1842–1912):

$$\mathrm{Re} = \frac{\rho \times A \times v^2}{\mu} = \frac{\rho \times l \times v}{\mu}$$

The length $l$ here refers to the characteristic length of an object—typically the shortest distance from front to back.

Now we're better placed to understand what a tiny swimming organism must deal with. Thanks to its dependence on both the length and velocity of a swimmer, the Reynolds number varies enormously across the size range of living things. It may reach 200 million for a whale as it goes about its business, whereas the equivalent figure for a bacterium may be only 0.0001. The tremendous implied change in the relative importance of inertial and viscous forces has an equally tremendous impact on locomotion. For most single-celled organisms, inertial forces are irrelevant—their world is wholly dominated by viscosity. The situation is often likened to swimming in syrup, although frankly this doesn't even begin to capture what these little creatures are up against: a human swimming in syrup at a leisurely speed of half a meter (a little under 20 inches) per second would experience a Reynolds number of about 50. In the desperately sticky world of the single-celled, much that we take for granted is ruled out. Coasting, for instance, is impossible: as soon as the engine is switched off, friction brings a tiny swimmer to a grinding halt very quickly indeed (the stopping distance for a bacterium is roughly equal to the width of a hydrogen atom). Lift, too, is out of bounds: in a low–Reynolds number realm, where fluid oozes around surfaces with ease, there's no near-surface fluid depletion and no all-important pressure difference perpendicular to the flow. For a similar reason, streamlining is worse than useless at this scale. Tapering the back end only helps when drag is a mainly inertial phenomenon. In the low-inertia world of tiny Reynolds numbers, the flow never separates no matter how blunt the trailing edge, so making the body more sleek and torpedo-like ends up adding drag on account of the extra surface area.

Thanks to the dwindling of lift at low Reynolds numbers, swimming microorganisms are limited to drag-based propulsion strategies, such as rowing. We've seen one such strategy already, in the cilia that some animals and algae use to get around. Cilia are used by various microbe factions that belong to the so-called eukaryote supergroup, all of which package their DNA in a membrane-bound nucleus. That distinguishes them from the prokaryotes—the older, smaller, and more primitive bacteria and the like whose DNA floats freely (sort of) within their cells. Plants, animals, and fungi are also eukaryotes, and so are more closely related to the nucleus-bearing microbes than any are to the prokaryotes. Cilia are common among the eukaryote microbes. However, animal-like ciliary fur-coat arrays are comparatively rare, being present in only one microbe lineage, the Ciliata (the group that includes the "slipper animalcule" *Paramecium*). Leeuwenhoek saw such cilia in action many times. They work like oars, with a power stroke (in which they're held stiffly erect and moved broadside to the flow), and a recovery stroke (when they're bent sharply and pulled forward parallel to the flow to reset). Because the drag is higher when the cilium moves broadside, thrust duly appears, although not as much as you might think. The general absence of flow separation at low Reynolds numbers means that an object's orientation has a relatively minor effect on drag. The saving grace of cilia is that they lie near the surface during the reset, and so encounter a slower-moving current (remember the velocity gradient)—that and the fact that they're always found in multiciliary arrays, and thus can make use of the wonderful metachronal waves.

The vast majority of single-celled swimmers have a somewhat different way of using their cilia, which are less numerous but longer than those in the Ciliata,* A typical eukaryote cilium, of which the tail of an animal sperm cell is a good example, undulates like an eel—or rather, the motion looks eel-like,

---

* These longer cilia are often referred to as flagella, but they are structurally identical to cilia, and so I employ the latter term here for both forms.

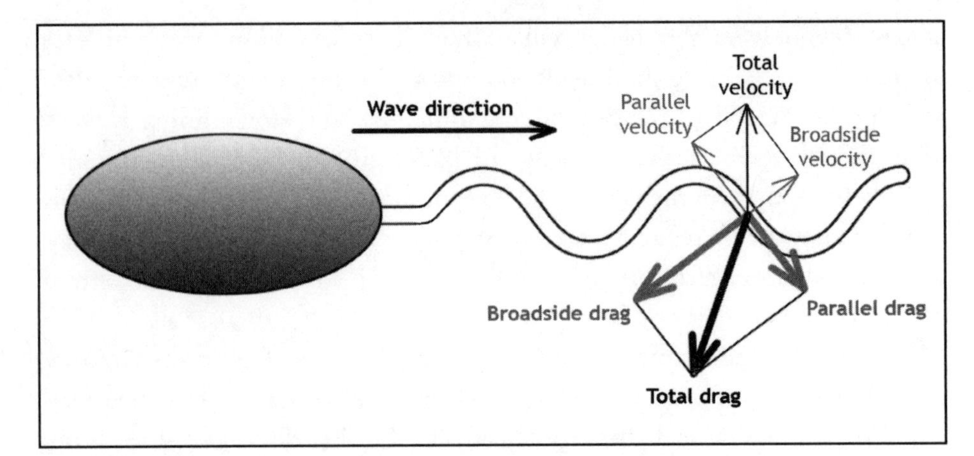


**9-2:** Undulatory ciliary swimming. As the wave passes from left to right along the cilium, any given segment of the structure oscillates vertically. The vertical velocity can be decomposed into components parallel and broadside to the segment, and because the latter component is resisted more strongly than the former, the total drag vector is tilted forward, providing thrust.

for the underlying physics is rather different, thanks to the much lower Reynolds number. An eel makes extensive use of lift (albeit directed horizontally), but as discussed, lift is off-limits for the very small.* Ciliary undulation is drag-based and, as for ciliary rowing, capitalizes on the fact that the drag of a cylinder moving broadside to the flow is higher than that of the same cylinder moving in parallel. As an undulatory wave passes down the cilium, any given short stretch moves obliquely from side to side. It therefore experiences both broadside and parallel drag forces, and because the former is larger than the latter, and assuming the ciliary wave passes backward with respect to the direction of travel, the overall drag vector tilts forward, giving thrust.

Now that we've seen how ciliary movements work, we might ask how these movements are generated, given the absence of muscles, which are of course the playthings of multicellular animals. It would be easy to dismiss this issue as unimportant in the grand scheme of things. But as we saw in Chapters 7 and 8, ciliary locomotion was used exclusively by the early members of the plant and animal kingdoms before the former abandoned locomotion and the latter evolved muscles, and thus had a key role in the inception

* I cannot help but wonder whether the increased availability of lift to undulatory swimmers as body size increases might have contributed to the grotesque enlargement of the vertebrates early in their evolutionary career.

of both groups. Furthermore, the distribution of these minute engines is sufficiently widespread to indicate that the last common ancestor of all modern eukaryotes had one. It's worth taking a moment to let that fact sink in. A cilium from your lungs and one from a single-celled photosynthetic alga are barely distinguishable from each other at the molecular level, despite the 2 billion years or so that have elapsed since we and it were the same species. It must be an astoundingly useful piece of equipment to have been preserved nearly unaltered for so long. If this were not evidence enough of its utility, every major branch of the eukaryote family tree carries organisms that use these ancient motors. We're clearly in the presence of something that defines a major chapter in the evolutionary history of locomotion—one that's well worth getting to know.

The framework of a cilium is a stereotypical arrangement of minute hollow tubes called microtubules, constructed from the protein tubulin, which extend the length of the structure. In the standard design, two run down the middle, surrounded by nine conjoined pairs called doublets.[†] This 9+2 arrangement is quite striking—almost artificial-looking—when viewed with an electron microscope. It was once thought that ciliary bending was caused by contraction of the microtubules on one side, but some careful microscopic work revealed otherwise. Under the contraction hypothesis, a cross-section taken near the tip of a bent cilium should be missing some microtubules near the inside edge. The opposite is the case—the *outside* edge is the one short of a few tubes. Peter Satir, now at the Albert Einstein College of Medicine, New York, realized back in the 1960s that this result would be explained if the tubules on the two sides hadn't changed length, but had slid past each other. The sliding causes bending because the tubules are restrained at their base—an effect that can be demonstrated if you have a grip-sealed plastic bag at hand. Cut the seal free from the bag, remove one end, and you'll have a pretty good model of a eukaryote cilium. Slide the two edges and the whole thing bends.

Around the time that Satir did his electron microscope work, the agent responsible for the sliding motion was in the process of being discovered, thanks to husband-and-wife team Barbara and Ian Gibbons at Harvard. Their approach was biochemical rather than microscopic: they soaked cilia

† The need to maintain contact between the outer and central doublets is what constrains the diameter of a cilium, which, as I mentioned in Chapter 7, is ultimately why large creatures can't use cilia as propulsive engines.

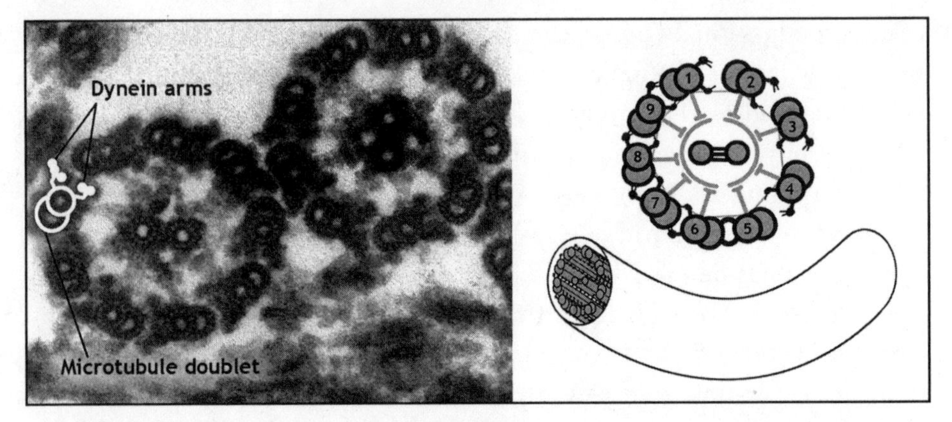


9-3: Eukaryotic cilia from the single-celled alga *Chlamydomonas*. The standard arrangement of nine microtubule doublets around a central pair is very clear under the electron microscope (left). When engaged, the dynein arms of a doublet slide the neighboring doublet toward the tip, but because all doublets are anchored at the base of the cilium, this action bends the structure. The dynein arms of doublets 1-4 bend the cilium in one direction, and those of doublets 6–9 bend it in the opposite direction (doublets 5 and 6 are permanently linked). The average spacing between the doublets, maintained by the "spokes," ensures that the dynein arms of only one side (doublets 6–9 in the diagram) can be engaged and active at any one time.

from sea urchin sperm (which are easy to obtain in large numbers without challenging anyone's modesty!) in various salt solutions, in an attempt to separate and identify the various protein components. One of these—later named dynein—was found to break down the molecule adenosine triphosphate (ATP). ATP is the standard biochemical energy currency of cells, so the fact that dynein could burn it suggested that this protein was the primary motor of a cilium—a suggestion later confirmed when cilia soaked in dynein-dissolving potassium chloride solutions were found to beat less vigorously. In an intact cilium, the dynein molecules are permanently bound to one of the tubules of each doublet like a series of tiny arms, all reaching toward the next-most-clockwise doublet (if looking from the tip). We now know that each acts as a lever, sequentially binding, tugging, and releasing the neighboring microtubule to bring about the observed sliding.

Which way a cilium bends depends on which doublets are engaged, as all the dynein arms tend to push the neighboring doublet tipward. In Figure 9-3, the dyneins on doublets 2–4 are responsible for bending the cilium up, whereas 7–9 bend it down. It's not known for sure how a cilium switches between these two states, or indeed how the beating parameters are adjusted to give undulatory or rowing patterns. The central pair of microtubules likely play a regulatory role, but as Charles Lindemann of Oakland University, Michigan, has persuasively argued, the basic ciliary bending sequence may have a purely geometric basis, the passive flattening and buckling of a bending cilium being largely responsible for controlling which doublets are engaged at any one time. Aside from its attractive simplicity, this theory helps to explain why nearby cilia coordinate their movements: the fluid flow set up by one cilium bends its neighbors and sets them off, too. Given the perennial significance of metachronal waves in the history of locomotion, that is no small matter.

## PROTEAN STRUGGLES

For all its locomotory usefulness and ancient provenance, the mark 9+2 cilium is too complex to have been the earliest eukaryote propulsive engine. Sadly, there are no living descendants of the preciliary eukaryote stock, but a number of later lineages have been so good as to lose their high-tech cilia and revert to a more basic locomotory system, giving us a glimpse of what it was to be a eukaryote in the days before the outboard motor evolved. As we'll see later, these Luddite eukaryotes also give us important information about where muscle comes from. The most famous of these throwback organisms are the wonderful shape-shifting amoebas, and it is to them that our story now turns.

Whereas ciliary propulsion is for navigating in water, amoeboid movement is usually used for crawling on a solid surface, although it works for swimming as well (just as cilia can be used for crawling, as in flatworms). The basic idea is to extend a blobby cellular protuberance, stick it to the ground, and then hitch the rest of the cell forward. The process is repeated to give a kind of walking movement, with the shifting, amorphous protuberances acting as rudimentary legs. For this reason, the blobs are called pseudopods (false legs), although they have no permanently defined identity—once they've done their bit, they're retracted back into the main body of the cell whence they came. Their formation is underlain by the construction of long flexible chains of a protein called actin, which build semirigid scaffolding-like networks when- and wherever a collection of molecules known as the Arp2/3 complex is locally activated. Importantly, an actin chain can only be added to at the front and dismantled from the rear. With appropriately deployed adhesives on the cell membrane, the whole actin trellis therefore shifts inexorably forward—it's rather like the prewheel technological strategy of rolling a stone slab on a set of logs, with the log at the back being moved to the front once it becomes unloaded. Thus is a pseudopod extended, and whether it grows or shrinks depends simply on the balance between the addition of actin molecules to the network at the front and their removal at the rear. The cellular skirt-hitching that happens at the back end of an amoeba is driven by the same kinds of actin manipulations, but it often makes use of another protein in the diverse myosin family. Like dynein, myosin is a walking motor, and can slide one actin filament relative to another just as dynein manipulates its microtubules. With sufficient overlap between the filaments toward the rear of the cell and those up front, a collection of myosin molecules will have little trouble pulling the back edge forward.

This amoeboid movement strategy was probably the earliest form of eukaryote locomotion, and it persisted in our ancestral lineage until the first fully functioning cilium appeared. At this point, the fortunes of our ancestors changed dramatically. Rather like the later transformation that would propel the chordates to global dominance, and indeed the much later origins of flight, the evolution of the cilium enabled a takeoff from the substrate, and thus unlocked a whole new world for our distant forebears: a world of three dimensions rather than two, within which the early members of the plant and animal kingdoms began their momentous careers. As with all major transitions, however, the construction of the new engine cannot have happened overnight. As the saying goes, what use is half a cilium?

As it happens, half a cilium is very useful indeed. It has long been known that many of the molecular components of a cilium can be found outside it, where they fulfil a whole host of nonlocomotory tasks. Microtubules, for instance, are present in all eukaryote cells, whether they have cilia or not. They typically radiate from an organizing center near the nucleus and act as the cell's primary scaffolding. They also fulfil a vital role if and when a cell divides by reorganizing themselves into the so-called spindle apparatus, which drags freshly replicated chromosomes apart. Like microtubules, dynein can also be found away from cilia, but rather than shifting one tubule with respect to another, their chief role in the wider cell is the transportation of cargo, using the microtubules as monorails.

With the components at hand, as it were, it's no longer difficult to envisage how the eukaryote cilium may have originated. As micropropulsion expert Gáspár Jékely has proposed, its evolution may have begun as a simple reorganization of a cell's microtubule scaffolding, with the organizing center building a long bundle of tubules in addition to the radiating fan to thrust an antenna-like process from the surface of the cell. This protocilium would not yet have been able to beat back and forth, but may have had a sensory role, giving the cell enhanced directional perception. If so, a constant traffic of motor proteins, such as dynein, would have been required to keep the antenna provisioned with receptor molecules. Alternatively, or additionally, the protocilium might have had a feeding function, with dynein being used to shuttle intercepted food particles back toward the cell. In either case, it would have taken only a slight evolutionary tweak for the shuttling dyneins to have started tugging on neighboring microtubules. The resulting movements, while crude at first, would have enhanced the sensory or grazing functions by causing the cilium to sweep the local area, or they may have enabled it to act as a permanent version of a pseudopod, helping the cell crawl along the substrate (as I said earlier, cilia are often used for surface gliding to this day). At this point, selection for improved locomotion could have taken over, gradually guiding the cilium toward its greater destiny.

## MOTION MEMORY

There's another, subtler consequence of the evolution of the cilium. My earlier allusion to the origin of chordates was perhaps a little disingenuous, for amoebas can paddle reasonably well with their pseudopods (the low Reynolds number makes this movement mode more effective than you might think). Suddenly, the cilium doesn't look so great, for it seems that we eukaryotes already had access to the 3-D world before it evolved. Of course, ciliary swimming might be more efficient or rapid than the amoeboid version, but there's another, more significant benefit relating to certain navigational difficulties at the single-celled scale.

When a large creature is working out where to go, it usually compares its current heading to the desired heading and adjusts its trajectory accordingly. This sort of thing could work, theoretically, for an amoeba. Say it was trying to find some tasty chemical: with a good spread of the relevant receptor molecules across the cell membrane and some internal molecular circuitry tying the local activation of a receptor to the construction of some actin scaffolding via Arp2/3, the little creature should be all set to thrust its pseudopods in whatever direction is tastiest. But the trouble with this strategy is that unless the concentration gradient of the chemical is very steep, the stimulation of the receptors on the near and far sides of an amoeba—separated by a mere fraction of a millimeter—will be nearly equal or even downright misleading. In other words, the chemical signal is liable to be completely swamped by the noise: random fluctuations in concentration that give no information at all about where a chemical has come from.

The usual remedy for a signal-to-noise ratio problem is to integrate the ambiguous inputs over time: after a while, the consistency of the signal becomes apparent against the randomly fluctuating background. The thing is, that's easy enough for animals, but we have a nervous system that can retain information for long periods of time. On the face of it, a single-celled amoeba has to rely instead on its receptor molecules, and while these can do a bit of input summation, their "memory" lasts no more than ten seconds, the precise time depending on how long the molecule binds to its receptor, how fast the participants in the relevant reactions diffuse, and so on.

So, how do amoebas work out where to go? Remarkably, they transcend the limitations of their small size by using their locomotion itself as a rudimentary kind of memory. Rather than waiting for a signal to arrive before setting off, most amoebas move continually, using a stereotypical zigzagging pattern of pseudopod extensions from the front end, with the same patches of the cell being used again and again as extension zones. The continual reuse of the same growth zones amounts to a directional memory, and it seems to happen because an old pseudopod's actin scaffolding takes a short while to be completely dismantled, and thus can be readily recycled. Rather like our left/right walking pattern, this pseudopodal zigzagging should take an amoeba in a straight line. If, however, a chemical gradient is present, the formation position of the pseudopods is slightly biased, such that they grow a little more to one side: either toward or away from the chemical source, depending on whether the substance is attractive or repulsive. The effect is only subtle, but over a number of zigzag pseudopod cycles, it's sufficient to steer the creature in the right direction.

The amoeba navigation strategy reveals an oft-neglected benefit of locomotion in the world of the small. For us animals, locomotion is an inherently responsive activity: we assess the environment and move accordingly. For the single-celled, however, locomotion isn't just what they do in response to environmental cues—it's how they perceive those cues in the first place. If they stop moving, they go blind. Which brings us to the real downside of the amoeboid way of life. Using a pseudopod-based locomotory memory to navigate by smell is one thing—finding one's way by light is an entirely different kettle of fish. A localized patch of photoreceptive molecules with some screening pigment to one side can give some information about the level of illumination, but it won't be much good at usefully communicating the direction of a light source to an organism unless it's used in a certain way. We've already seen how in Chapter 7: a spiral trajectory as used by *Platynereis* larvae should give a single-celled organism all the information it needs to navigate by light in three dimensions.

But there's an important condition: the relative positions of the eyespot and engines must be maintained, otherwise the navigational instructions will often be wildly inappropriate (imagine trying to drive a car in which the wheels sometimes turned with the steering wheel, sometimes against). The body must have a measure of morphological stability for 3-D phototaxis to be possible, and unfortunately, the most obvious attribute of amoebas is their distinct lack of morphological stability. There's no way around this: amoebas, and by extension the earliest eukaryotes, rely (or relied) on their shape-shifting skills to move. Until, that is, the cilium evolved. This is the hidden benefit that I spoke of earlier. The cilium may not be strictly necessary for swimming, but because it enables movement without requiring constant body deformations, it's a must for any swimmer that navigates by light. That might include the photosynthetic, the hunters of the photosynthetic, or those that merely want to head to the surface to cut down the number of dimensions for the purposes of, say, meeting a mate with minimum fuss. Once the cilium was in place, doing all this was merely a matter of evolving a simple eyespot and stabilizing the shape of the single-celled body.

## THE SHAPE OF THINGS TO COME

The easiest way to stabilize a cell is to surround it with a relatively rigid wall. Hence, the vast majority of light-guided eukaryotes are so endowed. They include plants, but also fungi (both kingdoms consisted entirely of single-celled wanderers in their early days), although they make their walls out of different stuff. Cellulose is the key structural component of plant cell walls, whereas fungi use chitin, otherwise seen in the exoskeletons of arthropods. This distinction indicates that the walls evolved independently in the two groups. The ciliates (i.e., the fur-coat-covered Ciliata) are among the few wall-less phototactic groups—they maintain cell shape with a jacket of vesicles that lie just beneath the cell membrane. Of course, the animal kingdom is another, but our ancestors took no specific measures to stabilize themselves until they became multicellular, which fits with their early predilection for a sedentary and largely light-irrelevant existence on the seafloor. That shape-holding measure, by the way, was none other than the origin of multicellularity itself: the old amoeboid shenanigans aren't possible for a cell that's surrounded by clinging clones.

Ironically, then, the earliest plants and fungi were considerably better at locomotion than were the first animals, thanks to the fact that animal cells never grew walls. But in a delicious twist, those self-same walls would later cause the plants and fungi to *abandon* locomotion. By staying wall-less, animals may have begun their career in the slow lane, but they also kept their locomotory options open, as is readily apparent today. For starters, many kinds of animal cells retain latent amoeboid capabilities that can be put to good use as the situation demands. White blood cells are obvious examples: their modus operandi when seeking out and destroying invading microorganisms—squirming their way past obstacles and sometimes wrapping themselves around the invader—is essentially identical to the way of life (and way of motion) of the very first eukaryotes. Individual migrating animal cells also play a vital role in wound healing: during this process, skin cells flanking the breach become partially amoeboid to migrate across the gap.

There is, unfortunately, a dark side to this free and easy locomotory individualism. When a rogue cell rebels against the usual multicellular checks and balances and starts dividing uncontrollably to build a tumor, the ability of cells to go amoeboid takes a very sinister turn. Once cancer cells break away from the primary tumor and worm their way into the circulatory system, they can wind up all over the body. The process is called metastasis, and it's responsible for the great majority of cancer deaths. For this reason, the various genetic switches that cause cells to transition to an amoeboid state—that is, to hark back to their ancient form—are currently under intense scrutiny, and hopes are high that these studies will give us more sophisticated weapons in our fight against this dreadful disease.

There's one more way in which the course of animal evolution has been influenced by our lack of walls and consequent access to the ancient amoeboid locomotory machinery. I mentioned earlier that some of the movements carried out by amoebas are mediated by an interaction between actin, which forms filaments, and myosin, a motor protein that can walk along those filaments as dynein shuttles along microtubules. In the straitjacket of a cilium, the tight organization of the dynein and microtubule components enabled a locomotory transformation. Well, the same thing happened with actin and myosin, and the result was that quintessential animal tissue: muscle.

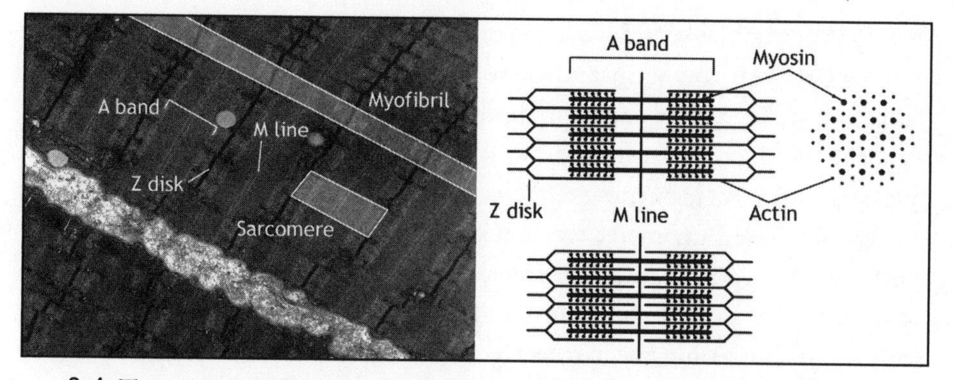


9-4: The structure of muscle. A muscle cell is packed with myofibrils (left), each consisting of a string of sarcomeres. The Z-disks at either end of a sarcomere are clearly visible under the electron microscope, as is the A band—the array of thick myosin filaments anchored at the M line. The diagrams to the right show part of a sarcomere in lateral view and cross-section in a relaxed (top) and contracted (bottom) state. Each thick filament is surrounded by thin actin filaments, and bristles with the globular heads of its constituent myosin molecules. When the muscle cell is activated, the myosin heads pull the neighboring actin filaments toward the M line by repeatedly binding to a filament, flexing, releasing, and resetting, using up one molecule of ATP for every cycle.

The organization of muscle cells has to be seen to be believed. Take one from any of our standard muscles and you'll find the same actin filaments that drive an amoeba's pseudopods, and similar myosin motors. But rather than forming an ill-defined network that's in a perpetually half-constructed state, the proteins are arranged with military precision. The principal organizational actin/myosin units—the battalions, if you will—are the myofibrils, each of which consists of a linear series of sarcomeres—the platoons. The sarcomeres are delineated at either end by pads of entwined proteins called Z-disks, with neighboring sarcomeres sharing the same disk. The Z-disks are where a sarcomere's actin filaments—all the same length and all lying absolutely parallel to each other—are anchored. Each myosin molecule consists of a bulbous head, a long tail, and an angled neck joining the two. Like their actin counterparts, but unlike amoeboid myosins, the myosin molecules of muscle are linked together into filaments, with the tails all lying parallel to each other and the heads sticking out to the sides. Whereas the actin filaments are found at either end of a sarcomere, the thicker myosin filaments sit in the middle, with enough overlap between the two filament types to ensure that every myosin head can reach an actin molecule if and when the signal to contract (usually a nerve impulse) is given.

The internal layout of a muscle cell is formidably complex, but the interactions between the actin and myosin molecules are similar to what goes on between the dyneins and microtubules of a cilium: the head of a myosin molecule binds to the nearest actin filament, flexes to pull on said filament, releases, and resets, burning a molecule of ATP in the process.* Zooming out to view one sarcomere, the collective action of all those tugging myosin heads pulls the Z-disks together, so increasing the overlap between the actin and myosin filaments (this is clearly visible under an electron microscope) and shortening the sarcomere. The contraction distance is tiny—only a few micrometers (millionths of a meter)—but because all the sarcomeres in a myofibril contract at the same time, their minute changes in length sum to give a reasonable range of motion for the whole muscle (unless prevented by an external force), which can then be further magnified using skeletal levers.

As is magnificently apparent, muscle allowed the animal kingdom to explore with a thoroughness that defies belief what locomotion really meant. In every sense, it let us boldly go where no one had gone before. But in biochemical terms, this passport to realms unknown required no great innovation. The basic apparatus had been around since the origin of the eukaryotes, and all we had to do to exploit it was stay wall-less, go multicellular, and do a bit of interior redecoration. This last step no doubt began in sponges, with their ultraslow contractions, but presently the cellular basis of these movements is unknown. However, the real turning point was the origin of the proteins of motion themselves: principally, the actins and myosins of the future muscle, and the dyneins and tubulins of future cilia. To understand where these eukaryote engines came from, we're going to have to enter the world of their forerunners—the prokaryotes.

## IN THE BEGINNING

The prokaryotes were the first cellular life-forms on Earth, but continue to thrive alongside their eukaryote cousins. Two large groups are known: the familiar bacteria and the superficially similar but biochemically distinct archaea, which were once thought to inhabit only extreme environments, such as hot springs or hyperconcentrated salt pans, but are now known to be

* This contraction process starts when an electrical impulse indirectly causes a molecular switch intertwined with the actin filaments—a protein called tropomyosin—to be pulled out of the way of myosin's binding sites.

every bit as ubiquitous as their bacterial cousins. In case you were wondering, genetic evidence indicates that it was the archaea that likely gave rise to the eukaryotes, although they couldn't have done so without the participation of bacteria (specifically, representatives of the alphaproteobacteria), which entered into a symbiotic association with the archaean eukaryotes-to-be and now function as the power stations of eukaryote cells—the mitochondria. In ecological terms, the prokaryotes are of immense significance—indeed, the biosphere wouldn't function without them: they're key players in all major biogeochemical cycles, by which such vital elements as nitrogen, phosphorus, and sulfur are recycled, and it's thanks to them (specifically, the cyanobacteria) that the planet became oxygenated. But that's another story. As far as our present purposes are concerned, the most interesting thing about the prokaryotes is that they were not only the first organisms, but the first to move. That puts them in a unique position to tell us about the most important event of all in the long history of locomotion—its origin.

But first, you need to know that likely prokaryote homologues of all the core molecular components of the eukaryote locomotory toolkit—actin, myosin, and the like—have now been found. However, they're not used to move prokaryotes from place to place; instead, they have important jobs when a cell grows or when it comes to divide: shuttling the replicated DNA to the two ends of the cell, for instance, or cinching its waist prior to splitting. Such intracellular machinery has obvious functional parallels with the eukaryotic cytoskeleton, so it might seem odd that the proteins don't have a locomotory role in prokaryotes. The reason is simple—the vast majority of these creatures have a cell wall (of an entirely different construction from any eukaryote wall), which rules out any kind of cytoskeleton-mediated, shape-shifting movements. For unknown reasons, the wall was lost in the protoeukaryote lineage—an event that not only opened the door to amoeboid locomotion, but also enabled the later construction of the cilium. Which is ironic, because many prokaryotes bear drag-based-swimming appendages—flagella—that in a broad sense work a bit like standard eukaryote cilia. We eukaryotes reinvented the wheel—a metaphor that has added significance in this context.

The bacterial flagellum has attracted some frankly unfair notoriety in recent years, being the onetime poster boy for the intelligent design movement: a dubious honor that owes much to the fact that it's a true rotary motor, with an axle at its base. Unlike the undulatory motion of the eukaryote cilium, the relatively inflexible bacterial version spins like a corkscrew, making it one of the few natural systems to functionally approach our much-lauded wheels. No wonder the intelligent design folk love it so much—if evolution hasn't managed to make a wheel yet, how could it have built a motorized corkscrew? Not that that's an argument they would ever use: an axle on a bacterial scale in an aquatic context is a world apart from the same in a macroscopic terrestrial wheel. On our scale, the necessary separation of axle and wheel is a deal-breaker (how can the wheel be nourished if its blood vessels keep getting tangled and snapped?), quite apart from the fact that wheels are only useful in a relatively flat or gently sloping environment with only scattered obstacles. These problems simply don't exist for swimming bacteria, and while their flagella have an undeniably impressive regularity, their mechanism of operation is not as seemingly unbiological as you might think. The axle's rotation is driven by flanking proteins (called MotAB complexes), and while the details have yet to be fully ironed out by the electron microscopists and X-ray crystallographers, it's thought that they spin the axle by reversibly changing shape to give a continual series of power strokes and recovery strokes. In functional terms, the system is therefore not *that* different from the motor-protein-driven movements we've already looked at. The only major difference is that the motion isn't driven by ATP, for once, but by the inward flow of hydrogen ions (i.e., protons) from outside the cell, where their concentration is higher, through the MotAB molecules. The necessary proton concentration gradient is maintained by pumps in the cell membrane.

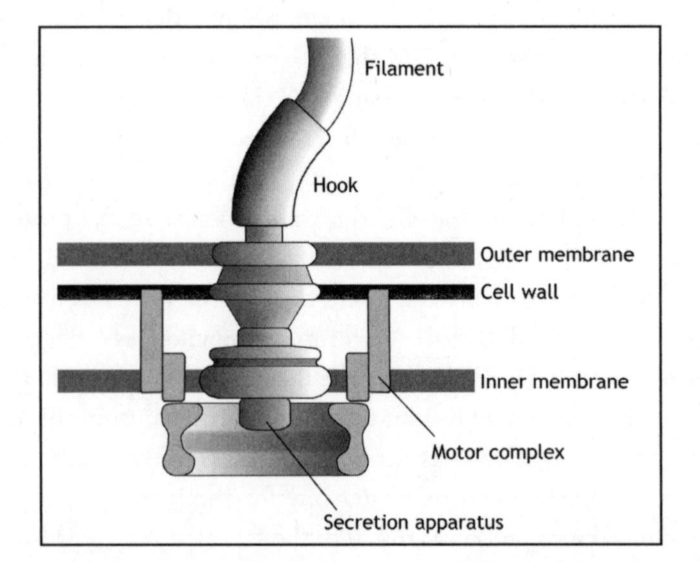


**9-5:** The base of a bacterial flagellum, with the lower ring cut away to show the underlying structure. Two sets of motor complex molecules are shown in section for clarity, but in reality these encircle the flagellar axle.

A typical bacterial flagellum has about twenty major protein constituents, with an additional twenty to thirty playing important supporting roles, such as constructing the device. These molecules variously contribute to the flagellar filament itself, the axle at its base, a hook that joins the two sections, the motor complexes, ring-shaped bearings, and a secretory unit with a set of molecular guides that ensure the right proteins get added at the right time. With so many interconnected components all fulfilling vital roles, flagellar function is quite sensitive to any perturbations of the hardware caused by mutations, which is handy for microbiologists: genetically screening locomotion-defective bacteria is a great way to dissect the flagellum's construction and operation. But, of course, it is this very sensitivity to change that creationists have jumped on, for if the loss of even one of these components compromises the workings of the flagellum, how could the structure have been built gradually by natural selection?

While this claim that the flagellum is irreducibly complex may seem sensible, it suffers from a major flaw: it rests on the assumption that the functions of biological structures are carved in stone, never changing. While this applies to most man-made devices, living machines can exhibit striking functional fluidity over evolutionary time, as we've seen again and again throughout the book. Our legs, remember, originally had an underwater propulsive function, and only later acquired their weight-bearing capabilities and complex joints. Our thumbs, now used for fine manipulation, evolved in the context of tree climbing. And to take an example from this very chapter, the eukaryote cilium components all had nonlocomotory functions before they were redeployed as a propulsive motor.

So, just because a stripped-back flagellum won't work as a locomotory structure doesn't mean it won't work at all. Spurred on by the intelligent design challenge, microbiologists have gone to great lengths to find homologues of flagellar components in other systems. Again and again they've succeeded, resulting in several detailed and wholly plausible models for the gradual evolution of the bacterial flagellum. Central to all scenarios is the realization that the filament is built by a secretory system. We can, as it happens, be more specific than that—many of the protein components of the flagellum can be found in the so-called type III secretory system, which is used by some pathogenic bacteria to build a kind of hypodermic needle for attacking eukaryotic cells. While the needle itself is not a homologue of the flagellar filament, an older type III system may have secreted similar threadlike appendages, or pili, to give the technical name, from which the flagella evolved.

Homologues of the MotAB drivers have also been identified: they constitute the so-called Tol-Pal system that stabilizes bacterial cell membranes. As this system has nothing to do with basic type III secretion, it must have been added to the apparatus later. The all-important rotary movement may have started with this admission of the Tol-Pal proteins to the protoflagellum club, although there's intriguing evidence that they may have taken over from an older motor. At the heart of the type III apparatus lies a collection of proteins that seem to be an early version of an F-ATPase—an ancient membrane-bound molecular machine that has the critically important job of making ATP using the energy stored in a proton gradient. Significantly, it carries out its actions by rotating, so maybe a proto-F-ATPase was the original protoflagellar rotor.

Many aspects of flagellar evolutionary scenarios are speculative at present. But we have at least demonstrated that natural selection *could* have built a bacterial flagellum gradually. If that were not enough, further evidence is available courtesy of the other prokaryotes—the archaea. Many of them have flagella, too, with the same rotary operational scheme and the same corkscrewing motion as their bacterial counterparts. But, on a molecular level, these flagella couldn't be more different. Everything from the protein constituents to the method of construction is unique to the archaea. This is clearly a case of convergent evolution: the selective pressures to improve swimming performance, coupled with the physical and biological constraints on life at a prokaryotic scale, have led to near-identical structural and functional solutions, but with different molecular building blocks. As to the specific nature of those constraints, it's doubtless significant that a prokaryote flagellum is necessarily thinner than a eukaryotic cilium—far too spindly to incorporate internal motor proteins—so the only way to get the sideways oblique movement that generates thrust is to build a rigid spiral and spin it.

The archaeal flagellum hasn't been studied as thoroughly as its bacterial analogue, but it too seems to be hiding an older secretion system at its core: not type III, but something resembling a more widespread system conveniently known as type II. Significantly, the simple threadlike pili that I mentioned earlier are secreted by type II–like nozzles, which lends support to the suggestion that such structures may have provided the template from which flagella were derived. Interestingly, pili often have a locomotory function themselves, but not swimming: they extend and retract to effect a kind of crawl. We may therefore need to add the independent origins of prokaryote flagella to the long list of 2-D to 3-D locomotory transitions involving takeoff from a substrate.

Every kind of prokaryote locomotory system we've looked at so far is underlain by a secretory system of one kind or another. In hindsight, this is hardly surprising—a locomotory appendage has to be extended from the cell somehow. But could we be looking at something more fundamental here? You see, pili-driven crawling and flagellar swimming aren't the only means of prokaryote transportation. Some entirely appendageless prokaryotes, including cyanobacteria, apparently propel themselves by using slime secretion at the back end as a rudimentary jet engine (the propulsive force is thought to derive from the expansion of the slime as it leaves the cell and absorbs water). The slime-extruding nozzles are similar to the external pores of type II and III secretion systems, raising the possibility, albeit tentatively, that slime-jetting was the ancient precursor of both pili-driven and flagellar mechanisms; it may even have been the very earliest propulsive technique. If so, the evolutionary precursors of flagella might have been locomotory all along.

The epic history of self-propulsion may thus have started with a few dribbles of snot from the backsides of early prokaryotes—hardly the most elegant of starting points. Neither was it cheap, with the continual loss of all that slime (which would help to explain later shifts to pili-driven crawling). Nevertheless, it's not hard to think of reasons why this ability was selected for in the early prokaryotes, despite the expense involved. Nutrients have never been evenly distributed in the oceans, so those organisms that were able to seek out the patches of soup would have gained a considerable selective advantage over those who were stuck in one place. But, of course, to find food a creature needs to do much more than simply move: it must move in the right direction, and that's not straightforward for something as tiny as a bacterium. Earlier we saw how single-celled eukaryotes are forced to implement a kind of locomotory memory to circumvent the dire signal-to-noise ratios that characterize their perceptual environment. At the prokaryote scale, even that won't work. With such a short distance between left- and right-hand sides, there's no way for a bacterium or archaean to tell which side smells more promising, even if integration is used to amplify the signal. Simply put, prokaryotes don't know where they're going. However, they're still remarkably good at finding what they need. Like eukaryotes, they use measurements in time to get around an inability to measure differences in space, but in a more extreme way. The tactics have been studied most thoroughly in the infamous *Escherischia* (*E.*) *coli*. This species has many flagella, all of which can switch between clockwise and counterclockwise spinning. When running counterclockwise, all the flagella unite into a single superflagellum, pushing the bacterium forward. If they switch to the clockwise mode, however, the bundle separates and the bacterium tumbles. With a switch back to counterclockwise spinning, the tiny creature sets off once more in a new, random direction. This simple alternation between runs and tumbles is the basis of *E. coli*'s navigational strategy. If it senses that the concentration of whatever it likes is increasing, it inhibits the flagellar switch and keeps going, but if the concentration is diminishing or staying the same, the switch activates and the bacterium tries its luck in a new direction. With this and other similar techniques, the earliest locomotory life-forms were able to take their first purposeful steps into the brave new world of self-propulsion. And the rest is history.

----------

As transitions go, the origin of locomotion may seem tame compared with other early-life landmarks, such as the origin of DNA or the genetic code. But don't let the apparent ease of the transformation fool you, for on the strength of its consequences, this was undoubtedly the most significant event in the history of life since it began. For a start, being able to move meant being able to find new sources of energy. There's now widespread (if not universal) agreement that life began on the ocean floor at hydrothermal vents, fueled by steep thermal and chemical gradients between the sea and the subsurface fluid. These days of course the bulk of the biosphere is supported instead by the Sun—a state of affairs that owes its existence to the development of photosynthesis in a few enterprising prokaryotes. Much is rightly made of the biochemical evolution that unlocked this vast, hitherto untapped energy source, but, of course, none of it would have happened unless those plucky prokaryotes were sunlit. Thus the origin of photosynthesis was necessarily preceded by an epic voyage from the depths of the ocean to the surface.

Finding new ways of charging the batteries was only the beginning of locomotion's influence on life's future direction. Once organisms started to move around, it was inevitable that they would run into each other, with every meeting bringing new selective opportunities: one creature might attempt to attack or run away from the other, or they might exchange genetic material or cooperate in some other way. In other words, it was with the onset of locomotion that ecosystems began to take on a modern flavor, with producers and consumers, predators and prey, sexual relations and symbioses. And evolution became as much about organisms adapting to one another as to their environment. To cut it in this brave new world, living things had to continually find new and improved ways to defend or arm themselves, to hide from their enemies or seek their friends and victims, and, of course, to race, chase, or escape from each other. Thus was the stage set for a relentless locomotory arms race whose various fruits—spinning, undulating, crawling, rowing, walking, jetting, running, climbing, gliding, flying—could fill a book (and indeed they just have). On a more subtle note, the origin of locomotion brought with it a certain responsibility. Being able to move is of little use if you can't make an informed decision about where to go, even if that decision is, as in prokaryotes, a simple binary choice between holding your course if things are going well or randomly changing direction if they're not. To make anything of their powers of movement, organisms have to gather information from their environment, filter the useful from the useless, prioritize the remainder, and respond appropriately. This is intelligence in its most basic form.

Finally, and most crucially, it's only thanks to locomotion that life had any longevity. Motile creatures can avoid local extinction events, which would have been frequent if life really did start at hydrothermal vents: being perched on an oceanic crust conveyor belt, these structures have a limited shelf life, because once they've moved away from the spreading zone, they cool and shut down. Any nonmotile creature still on board at the time would have gone down with the ship—eventually right down, as the oceanic crust was subducted into the mantle.

While we're rightly fascinated by the emergence of the core biochemistry of life—DNA, RNA, proteins, cell membranes, and the metabolic and reproductive processes they embody—a biosphere that stopped at this point would barely justify the name. It wasn't until locomotion came on the scene that the living world came of age and became more than biochemistry alone. If self-propulsion had never evolved, life would be nothing more than a few scattered and short-lived patches of unusually complex chemistry, running inconsequentially on the ocean floor of an otherwise dead planet.

## 10

# Locomotive Souls

### In which we learn how motion is good for the mind

Now shall I walk
Or shall I ride?
"Ride," Pleasure said;
"Walk," Joy replied.

—W. H. Davies, *The Best Friend*

The first sign that something's wrong is an uncharacteristic solitary wandering. For most insects this would be perfectly normal, but for a worker carpenter ant, such antisocial behavior is bizarre in the extreme. Taking leave of its colony high in the rain forest canopy, it strikes out from the well-beaten foraging trails and meanders aimlessly from leaf to leaf. Suddenly, its body is racked with convulsions and it loses its grip, plummeting to the forest floor far below. Ants being resilient creatures, it picks itself up and tries to return home, but it can get no higher than a foot off the ground before the spasms strike again. The futile cycle continues for a while, until about noon, when the insect seeks out the underside of a leaf of whatever plant it's been struggling to climb, and bites down hard on the midrib. That will turn out to be its last act, for the fungus that's been growing inside its body has broken down its jaw-opening muscles. It takes anything up to six hours for the ant to die, and a further two days or so before the reason for its demise becomes apparent, when the parasite's spore-packed fruiting body emerges from the back of its neck. Once mature, the spores are dropped, ready to infect another unfortunate worker, and the cycle begins again.

This ghastly fungus goes by the name of *Ophiocordyceps unilateralis*, one of a large group of arthropod parasites, specialists all of particular host species. The infection cycle plays out in much the same way for each, but what sets the ant-killer apart is the extraordinary control the fungus gains over its host in the last few hours of its life. Almost all dead zombie ants (as they're popularly known) can be found about 10 inches off the forest floor, with a bias toward the north-northwest side of plants. While this might look like some kind of macabre artistry, David Hughes and his research group at Penn State University and their collaborators have found that this fussiness on the part of a zombie ant is enormously important for the survival of the fungus. If the ant were to climb too far before dying, or end up on the upper surface of a leaf instead of the shady underside, the fungus would get too hot and dry to develop properly. Conversely, if the ant were to expire at ground level it would soon be picked off by scavengers or washed away by the rain. To successfully reproduce, the fungus must find the Goldilocks zone between these lethal extremes, and to do that it must take complete control of the ant's locomotory system—body and mind. How it does so is still a mystery, although Hughes is beginning to uncover pieces of the jigsaw puzzle: the atrophy of the jaw muscles is clearly part of the parasite's strategy, as are the spasms that interfere with an ant's ascent. However it does it, the infiltration of the ant's being is so successful that the insect effectively becomes an extension of the fungus, existing only to maximize the propagation of the parasite's genes. In a sense, and despite the inherently sedentary nature of its fungal kin, *Ophiocordyceps unilateralis* is an organism with neurally guided locomotory powers, at least while it's driving its host. It's as if a jellyfish were sitting at the controls of a Harrier Jump Jet.

An *Ophiocordyceps*-infected zombie ant is a particularly impressive if ghoulish example of an extended phenotype. For the fungal genes to succeed, they must employ all kinds of subtle biochemical tricks to make the ant's body work to their advantage. It all sounds shockingly underhanded. But if you think about it, this sly manipulation isn't that far removed from the relationship between an ant's body and its *own* genes, thanks to its locomotory ability. Motile organisms can respond to change much more quickly than their genes, so those genes, if they are to have any measure of success,

10-1: An unfortunate carpenter ant falls prey to the fungus *Ophiocordyceps unilateralis.*

must take an indirect route to high Darwinian fitness: by building their host organism's sensory and propulsive systems in such a way that the creature acts in their best interests, speaking metaphorically.* Of course, the needs of the organism's own genes will usually (but not always) be served better if the organism itself survives, whereas a zombie ant's fungal genes need it to die, but the essential idea is the same.

Early on in the history of locomotion, genes' needs could be met very simply by, say, implementing a molecular switch to cause a prokaryote-grade host to keep going if things smelled good or change direction if not. With the rise of the animal kingdom and its trademark nerves, however, the situation became anything but simple; in fact, the nervous systems and the behaviors they controlled became so complex that the genes began to take something of a backseat even during the construction stage, by letting the nerve networks reshape themselves through learning. Then came the ultimate challenge when, for reasons unknown but probably related to learning, consciousness arose. It's one thing to control a mindless automaton—how can the genes get what they metaphorically desire when their host is genuinely aware, with real desires?

* To streamline our explanations, we evolutionary biologists often speak of genes as if they have motives. Needless to say, genes don't really "want" anything: they're just chemicals doing what chemicals do. Only conscious entities can actually desire.

In a sense, I have already answered my question: emotions are the tools—inherent in the neuronal wiring of the brain and its various neurotransmitters—by which genes persuade their conscious host to do what's best for them. Those behaviors that cause a reduction in lifetime reproductive success are punished with pain and fear, whereas anything that's likely to increase fitness is rewarded with excitement, satisfaction, and joy. As locomotion itself scores high on the fitness-enhancing scale, we would expect the very act of movement to be enjoyable; so, too, any training that improves propulsive efficiency. We thus have an evolutionary explanation for the phenomenon that goes by the delightful German word *Funktionslust*, which roughly translates as "the pleasure one takes in doing something well." It's the ultimate tool by which genes get conscious species to use their locomotory machinery as effectively as possible.

And, yes, I do mean species in plural. The scientific consensus is now aligned with what many of us were sure of all along—that *Homo sapiens* is but one of many conscious animals, all of whom can feel fear, pain, and joy.* We would therefore expect these creatures to feel a certain *Funktionslust* when practicing their locomotory skills, particularly if their survival or reproductive success is especially dependent on such skills. It's not hard to find likely examples. Many are the times that I've watched gulls flying beside the cliffs near my home town, apparently doing nothing more than playing with the deflected air currents: they hang motionless for a while before plunging at breakneck speed toward the sea far below, then use their momentum to climb back up again. While the ultimate genetic "aim" of these antics may be to hone and maintain their aerial skills, I have no doubt that on a conscious level they're just swooping around for sheer enjoyment. Wouldn't you?

There's little we can say at the moment about where on the tree of life the undoubtedly blurred line (or lines) should be drawn between the conscious and the unconscious. We humans, however, most certainly fall in the former category. Given how important locomotion has been to our ultimate evolutionary success, we have every reason to expect that the emotional bribery used by genes to get their hosts moving, and moving well, will be fully

* There's a curious undercurrent in some scientific circles that rejects the attribution of joy and pleasure to nonhumans, while having no problem at all with nonhuman pain and fear. As I see it, it makes no sense for natural selection to bring about one side of the emotional persuasion toolkit, then wait patiently for the appearance of some mysterious attribute borne only by humans before adding the other side. It all smacks of a last-ditch attempt to draw a line between us and the rest of the animal kingdom.

evident in us. In this last chapter of our journey we'll see that this is indeed the case: just as our body makes no sense unless appraised in the light of locomotion, so, too, does our mind make little sense unless viewed in the light of our genetically induced desire for locomotion. But there's a snag. The recent explosive expansion of our cultural repertoire has caused our locomotory drive to become expressed in all manner of curious guises, some of which are undermining the whole delicately balanced enterprise, putting our minds and bodies in terrible danger. More on this later. First, however, we're going to look at the very beginnings of our love affair with locomotion.

## BABY STEPS

The magical moment when we take our first few teetering steps is rightly regarded as one of our most important early achievements. It takes on average about a year of growth and training for our strength and skill to build sufficiently to walk unaided, and several further months before we're able to do so fluently, although there's enormous variation around this average figure. Perhaps more surprisingly, the *way* in which this skill is acquired is similarly variable. The "classic" view (based on a Western bias) is that babies pass through a hands-and-knees crawling stage before rearing up on their hindlimbs. However, the crawling stage may be skipped completely; babies may practice all manner of alternative modes before they walk, including bottom shuffling, inchworming, commando crawling, upright-legged bear crawling, and log-rolling. Basically, before they're strong or skillful enough to walk, babies will happily try out all kinds of locomotory techniques.

Given this tremendous variation, it might seem surprising that, barring injury or disease, we all end up converging on the same locomotory pattern. Why, for instance, don't any of us stop at crawling? It once served us well—why abandon it in favor of a less stable bipedal gait at which we initially have no expertise? This is one of the many puzzles that baby locomotion expert Karen Adolph and her group at New York University have started to unravel. In one study they closely observed a large number of infants—some crawlers, some new or experienced walkers—ranging in age from twelve to twenty months, going about their business in a playroomlike lab. There was no coercion whatsoever—the children moved as and when they desired. Adolph's team found that, unsurprisingly, novice walkers fell more often than expert crawlers. However, they also spent a lot more time in motion, and covered far greater distances—300 meters (almost 1,000 feet) in an hour-long session, whereas crawlers managed only 100. That meant that the number of falls per unit distance traveled was *lower* in novice walkers than in crawlers.

It thus appears that infants switch to walking as soon as they can, simply because, in spite of appearances, it's a more effective method of locomotion from the very first step, allowing them to move more quickly and with less effort. That the effectiveness of locomotion correlates with the time spent doing it indicates that the motivation to move increases as infants get better at it. These trends continue once they're upright. The first steps are short and stuttering, with quick swings to minimize the single-leg support phase. The legs are placed far apart to afford a larger polygon of support, the hips and knees are bent, chimp style, and the extensor and flexor muscles show a marked tendency to contract in sync, which improves joint stability but costs energy. This labor-intensive grind soon gives way to a more fluent progression, as a toddler learns to use gravity instead of fighting it, alternates the flexors and extensors rather than firing them together, and generally comes to trust his or her own body as muscle activation sequences become habitual, greatly easing the attentional load. The better infants get at moving, the more they appear to enjoy it, the more they enjoy moving the more they move, and the more they move the better they get—it's a wonderful positive feedback loop of *Funktionslust* that eventually gets us using our anatomical gear to the best of our abilities. Because human bodies are fairly similar, it's therefore no surprise that we all end up moving in the same way.

As important as our movement skills are, becoming locomotory has an impact on our mind that goes far beyond simply learning how to put one foot in front of the other. This first became clear in various landmark psychological studies carried out by the late Margaret Mahler, Jean Piaget, Eleanor "Jackie" Gibson, and others in the latter half of the twentieth century. Mahler in particular argued that the onset of self-propulsion marks our psychological birth, so profoundly does it transform our relationship with our surroundings and other people—a claim reinforced by many more recent studies.* Most obviously, our wider environment stops being a mere

* These include investigations overseen by Linda Acredolo at the University of California, Davis; Melissa Clearfield at Whitman College, Washington; Joseph Campos at the University of California, Berkeley; David Anderson at San Francisco State University, Bennett Bertenthal at Indiana University, Gwen Gustafson at the University of Connecticut, Michael Tomasello at the Max Planck Institute for Evolutionary Anthropology, Leipzig; and Karen Adolph at NYU. These studies supplied the information in this and the next paragraph.

backdrop and becomes something with which we can interact. As distant objects become relevant, we begin to direct our attention to far space. We soon discover, however, that it takes time to reach faraway locations, and so we learn to tolerate a delay between having an intention and experiencing its realization. Our working memory is concordantly honed to keep our deferred goals in mind. As we move around, we also learn a whole catalog of spatial rules. We find that objects seem to grow as they're approached, but that this is an illusion—indeed, we learn to use such clues to help us judge distance. We work out that the location of items is fixed not with respect to ourselves but to an external reference frame (a toy on the left ends up on the right if we move past it and turn around). These experiences greatly improve our performance at spatial problem-solving tasks (finding hidden objects, for instance), and our geographical memory.

Our acquisition of locomotion also has a profound influence on our social lives, for it allows us to leave the protective cocoon of our mother's immediate influence and begin to forge our own identity. But we stay in touch, quickly learning that we can look to Mom for information if we're unsure of anything. With our accumulating experience of executing goal-directed actions, we begin to work out how to read intentions into the actions of others, and how to influence these intentions: for instance, toddlers become adept at directing their mother's attention to items of interest. Movement thus seems to facilitate our development of a theory of mind—the attribution of mental states to oneself or others regarded (perhaps wrongly) as a quintessentially human trait.

Now, it's possible to overstate the influence of locomotion. The attainment of self-propulsive skills may be delayed or even prevented by congenital diseases, but as long as those diseases don't have a direct psychological impact, social and physical reasoning skills usually end up developing normally, thanks perhaps to the compensating effects of language (an argument put forward by James Rivière at the University of Rouen). Furthermore, some workers, such as Emily Bushnell at Tufts University, think that we underestimate what a passively moved baby learns about the world's physical rules. That said, it can't be denied that our early wanderings act as a kind of psychological scaffolding, even if there are backups, broadening our awareness of the world. But what happens to this scaffolding once we reach adulthood? Our movements obviously remain dependent on our mind for guidance, but to what extent does our mind continue to depend on our movements?

## THE NATURAL HIGH

We humans have been dealt a singularly strange hand by natural selection. Just a few million years ago, our ancestors were run-of-the-mill primates, hanging out in trees and eating fruit. A mere few hundred thousand generations later, our forebears—terrestrial Johnny-come-latelies in the grand scheme of things—were hunting creatures across the open savannah whose locomotory performance had been honed to near perfection by a protracted arms race with pursuit carnivores. On paper our ancestors had no chance, but as we saw in Chapter 1, their superior thermal physiology—an unexpected bonus of bipedalism—allowed them to use their locomotion itself as a weapon; to harry their four-legged prey to exhaustion. However, persistence hunting relies on more than sweat alone. Running an antelope to death can take hours, and for much of that time the prey won't even be visible, having galloped off into the distance at the first sign of trouble. Maintaining contact takes formidable tracking skills and a will of iron. That means that if there was ever an animal that needed to get pleasure out of locomotion, it's us. Without the neural buzz to mask the physiological stress of continuous high-level exercise, our ancestors would surely have given up long before their targets collapsed.

For most of us, of course, our hunting days are long done, and you might therefore expect our psychological adaptations for that way of life to have faded away by now. But on an evolutionary timescale, ten thousand years (the time that's passed since the development of agriculture) isn't that long, and all the while our culture has been buffering us against the transformative influence of natural selection. That's why I can comfortably live at a latitude of 52 degrees north despite lacking a fur coat. It's also why we may become terrified of snakes at the drop of a hat, despite the conspicuous lack of deadly snakes in many of the environments we call home. In Africa it's a different story: there, a phobia of snakes can save your life. If this aspect of our psychology has changed so little since we were hunter-gatherers, could the same be true of our attitude to running?

Some would answer with an emphatic yes. Despite the fact that vanishingly few of us need to chase our food these days, making the activity in rational terms a complete waste of energy, running can give us great pleasure. I speak of the legendary runner's high—the heightened sense of well-being and euphoria that endurance running can engender. These feelings used to be attributed to an increased concentration of endorphins, but we now know that that distinction belongs instead to the so-called endocannabinoids, or eCBs (named for their resemblance to the active ingredient of cannabis).

These are powerful painkillers, released during long-distance running, that also stimulate the secretion of the neurotransmitter dopamine from various neuron populations in the brain. Dopamine is popularly conceived as the brain's pleasure-inducing reward chemical, and while there's more to the neurotransmitter than that, as we'll see shortly, the general notion is supported by the fact that many recreational drugs, including cannabis, of course, but also nicotine, heroin, and cocaine, either stimulate the release of dopamine, inhibit its absorption, or mimic its actions.

All of that raises a perplexing question: if running feels so good—equivalent to taking illicit drugs without the nasty physiological, psychological, or legal side effects—why aren't we doing more of it? The generally inadequate level of physical activity in many human societies is a major concern at the moment. According to the World Health Organization, obesity has nearly doubled since 1980, and even back in 2008 about 35 percent of the global adult population were overweight, and therefore at increased risk of cardiovascular diseases, diabetes, and certain cancers. Obviously, our energy-rich diet is partly to blame, particularly in the West (another relic of our hunter-gatherer past—we're programmed to crave calories), but that wouldn't be such a problem if we had a greater collective will to run off those calories.

One clue to the resolution of this paradox is that only certain kinds of exercise induce the eCB high. We must remember the context in which this particular neural reward system evolved. Our ancestors needed it to persuade them to chase prey for hours on end, which, as I said earlier, would have put their systems under a fair amount of stress. The function of the eCBs was to suppress the sensation of these unpleasant side effects in favor of the long-term gain. Simply going for a walk doesn't stress the body nearly enough to switch on the eCB system. This catch-22 doesn't just apply to humans, by the way. David Raichlen of the University of Arizona and his colleagues have shown that dogs—the ultimate running junkies—are in the same bind. The concentration of eCBs in the bloodstream of his canine test subjects went up markedly after a thirty-minute bout of endurance running, but showed a significant *drop* after walking (on a treadmill, though, which might be significant). Interestingly, Raichlen's team also found that eCB levels in ferrets showed no significant change after walking or running, which emphasizes how important it is to keep the adaptive context in mind when thinking about the psychology of exercise. Polecats (of which ferrets are the domesticated descendants) have no use for endurance running in their day-to-day lives, and so have no need for a human/wolf-style eCB system.

What about the other end of the exercise spectrum? If physiological stress is key, do the more extreme demands of sprinting induce the high? I'm afraid not: that would be too dangerous. When running flat out, the demand for oxygen by our muscles outstrips the circulatory supply, forcing them to respire anaerobically. We can't keep this up for long—anaerobic metabolism generates lactic acid, which soon accumulates to dangerous levels if we don't stop and deal with it. If we were to successfully suppress the unpleasant sensations associated with such a buildup, we'd run ourselves to death through acute acidosis. That doesn't mean that high-intensity exercise isn't useful in health terms—short bouts are surprisingly effective at burning calories—but it doesn't feel very nice (in an emergency situation, adrenaline has the job of temporarily overriding this unpleasantness). Here's the rub. Unless our aerobic fitness is quite high, we won't necessarily have to run at full tilt to feel the lactic acid pinch. A low cardiovascular capacity might have trouble supporting a reasonably prolonged jog, and if so, our protective mechanisms may simply prevent us reaching the level at which the eCB system is activated. Until cardiovascular fitness improves, the eCB high will remain out of reach, and endurance running will just feel grim.

The take-home message is that if we want to experience some good old-fashioned locomotory joy, our activity levels need to be as similar as possible to those of our ancestors. For many of us, that might sound a bit depressing, particularly those who know anything about today's elite hunter-gatherer cultures. The legendary Tarahumara people of northern Mexico, for instance, can run 100-odd miles in a single session. But don't panic! Such seemingly superhuman feats should be considered extreme outliers, even by ancestral standards. James O'Keefe at the University of Missouri and colleagues recently carried out an in-depth analysis of traditional hunter-gatherer lifestyles, and found that marathon-length endurance runs usually go way beyond the genetic call of duty.* Typical estimated distances covered daily by hunter-gatherers are a surprisingly modest 3 to 10 miles, although there's a great deal of variation around these average values from day to day and from population to population.

Indeed, nice as it is, you don't need to achieve the runner's high to get both physical and psychological benefits of exercise: the eCB buzz isn't

* Marathons can even be bad for you. Aside from the merry havoc played with knees, hips, and backs, a number of studies have found that regular prolonged running can scar the heart.

some kind of exercise litmus test. After all, we used to be hunter-*gatherers*, not just hunters. Now, gathering can be strenuous, involving long, energy-sapping walks, but the intensity of effort is much lower than in a protracted hunt. Without locomotion-induced stress, there's no need to activate the alarm-overriding eCB system; that doesn't, however, make walking a joyless trudge. The bottom line is that the runner's high is just one aspect of locomotory pleasure. Other kinds of locomotion are fun, too—they have to be, otherwise conscious animals wouldn't be bothered to go anywhere. Remember, ferrets don't use eCBs, but they're hardly lacking in locomotory drive. The question is, without the dopamine high that the eCBs activate, what's in it for them, or us?

## THE TRUTH ABOUT DOPAMINE

Dopamine, as I said earlier, is traditionally characterized as a reward neurotransmitter in the brain, responsible for the sensation of pleasure. There are some important implications of this notion. Addictive behaviors, for instance, might be predicted to be caused by excessive dopamine function, whether this be achieved by extra release of the chemical, a failure to reabsorb it, or any other mechanism with similar ends: the intense pleasure induced by the dopamine excess keeps us wanting more. Experiments on rats carried out by Canadian psychologists James Olds and Peter Milner in the 1950s seemed to bear out this hypothesis. They implanted electrodes deep in the rats' brains that delivered a brief impulse whenever the rats pushed a lever. Which they did, over and over again. In fact, they self-stimulated up to two thousand times an hour, putting off feeding, drinking, sex, and pretty much any other behavior you could think of, just so they could keep pressing that lever. The region of the brain whose stimulation was responsible for this extreme addiction—the septum—is known to receive synaptic input from dopamine-secreting neurons, immediately implicating dopamine as a pleasure chemical. The later findings that several recreational drugs enhance dopamine function seemed to confirm this hypothesis. However, there were some troubling inconsistencies. For example, lever-addicted rats showed none of their usual facial pleasure signs, calling to mind a long-term addict who's developed drug tolerance.

A growing number of scientists now think that we've been getting dopamine wrong all these years. Among those spearheading the revisionist thinking are Jeff Beeler and Xiaoxi Zhuang at the University of Chicago. They work with mutant mice that lack the molecular mop that reabsorbs dopamine once it's been released at a synapse. In effect, such mice are permanently coked up. To see how the extra dopamine affected their behavior, Beeler and Zhuang devised an ingenious variant of a simple lever-pushing task. They gave the mice two levers to press. Pressing either led to the delivery of food, but one was more expensive than the other in terms of the required effort—it had to be pressed many more times before yielding the goodies. Every so often the lever protocol was switched, such that the cheap lever became the expensive lever and vice versa. Both the mutant mice and a set of normal mice quickly learned which lever was which, and cottoned on to the switch whenever it happened. However, in the stable periods between switches the two kinds of mice behaved very differently. The normal mice went for the cheap lever almost every time, but the dopamine-enhanced mice were quite happy to work harder for their lunch, despite there being no difference in the quality of the pellets.

These and other experiments have led Beeler to conclude that dopamine doesn't simply elicit pleasure whenever it's released; its ultimate role is more nuanced. Specifically, by emotional manipulation it seems to control the amount of effort that an animal is willing to expend to get what it wants. Moreover, the dopamine level dictates whether it falls back on its previous learning or tries something new to find out more about its environment. In a low-dopamine state, an animal is a lethargic creature of habit, unwilling to do anything more than the bare minimum; in a high-dopamine state, it's active, attentive, and eager to explore its world. The critical determinant of which of these two courses of action is appropriate is the availability of energy. If resources are plentiful, an animal will lose a lot more by doing nothing than if it spends those resources on acquiring new information or skills. If, on the other hand, resources are scarce, it won't be able to afford such exploratory extravagance. The dopamine system's main job is to ensure that an animal calibrates its activities accordingly—it's the key interface between the creature's nutritional state and its locomotory behavior.

That, at least, is the idea. Many details of this new dopamine hypothesis, if I can call it that, have yet to be rigorously tested, and there are bound to be complications. For instance, dopamine is known to help build habits in some situations, moving the neuron-firing patterns associated with newly learned behaviors from the apparently consciousness-invested motor cortex to the underlying basal ganglia, where the complex motor patterns can be accessed with fewer attentional demands (this is what we refer to when we speak of muscle memory). That role seems to conflict with the idea that dopamine promotes exploration. However, this impression may have much to do with a generally pejorative aura that hangs around the word *habit*, which tends to call to mind bad habits, such as smoking or nose-picking. If the process is reframed as skill mastery or training, the involvement of dopamine seems wholly appropriate. Viewed in this light, habit forming is exploration that commits.

That the dopamine system turns out to have hidden functional facets is in hindsight hardly surprising. There are tens of thousands of dopamine-secreting neurons in the brain and several different kinds of dopamine receptor: of course the neurotransmitter has multiple roles. To give another example, some dopamine circuits are involved with low-level motor control rather than abstract goal-chasing—these are the neurons hit by Parkinson's disease, hence the uncontrollable shakes that can be ameliorated by dopamine-mimicking drugs. The complexity partly explains why recreational drugs do such psychological harm—our narcotics hijack some aspects of the system but not others, putting the mind out of kilter.

Notwithstanding the intricacies of the dopamine system, it's possible to identify a common functional theme, much as we can with adrenaline, whose multiplicity of functions collectively ready an animal for fight or flight. Dopamine's various interests—motivation, motor control, attention, and motor learning—are all concerned with ensuring that an animal makes the most of its motor apparatus—its locomotory apparatus first and foremost—within limits set by its resource status. The neurotransmitter has had this role for a very long time: dopamine-secreting neurons are present in invertebrate brains, and a number of studies of learning in our spineless cousins, particularly fruit flies and honeybees, have uncovered many of the functions seen in mammals. For instance, dopamine enhances locomotion in these species. The system must therefore be at least as old as Urbi, and likely had the same basic role.

## THE LOCOMOTORY GOOD LIFE

Now, few people if any would argue that Urbi was conscious. While it may be worth entertaining that possibility once we have a firmer handle on what consciousness is, we can safely assume for now that awareness arose at a considerably later date. Dopamine's much-loved pleasure-giving role likely developed hand in hand, for the purposes of persuading more complex, sentient life-forms to keep doing what had previously been done automatically to evident success in their simpler forebears. Or, to put it another way, from the moment consciousness was on the scene, dopamine-induced pleasure diminished an organism's emotional sensitivity to the costs of locomotion whenever they could be tolerated.

Going back to our earlier question, the dopamine-related sensations of the runner's high must represent a late addition to the overall joy-giving role of the dopamine system. *Any* nonemergency locomotory effort should on the face of it be pleasurable as long as it doesn't cause dangerous departures from physiological balance: the gearing up of certain parts of the dopamine system by eCBs merely gives an extra burst to override the stress signals associated with endurance running. For instance, we can be quite sure that the exploratory locomotion of babies is strongly encouraged, and doubtless rewarded, by the dopamine system.

But are we overestimating dopamine's emotional power? Most of us know that walking *can* be enjoyable, but we don't experience a warm, fuzzy feeling every time we take a few steps. As always, we need to appreciate the ancestral context of our psychological design. Dopamine doesn't just induce locomotion because we've got nothing better to do. What matters is exploration: using locomotion to learn more about ourselves (by mastering new skills) and about our environment. Dopamine will have little to do with a simple trudge down a road we've walked hundreds of times, because it's an activity that requires almost no conscious engagement. Contrast that with, say, an extreme sport like surfing, skiing, parkour (free-running), or skateboarding. As well as the obvious need to master new motor skills, these activities require an intense level of concentration, as the slightest error of judgment or failure to pick up on a vital cue could lead to broken limbs or worse. Such complete immersion in a task is the hallmark of the much-vaunted flow state—a delightful (if seemingly paradoxical) combination of calm and excitement that we experience when completely absorbed by a challenging activity. It's quintessential *Funktionslust*, and has dopamine written all over it.

For reasons of nature or nurture, or simply age, extreme sports aren't everyone's cup of tea. Fortunately, we don't necessarily have to go to such lengths to experience a bit of locomotory joy. The adrenaline rush of extreme sports—usually caused by the high accelerations that occur when playing games with gravity*—is a bonus. In theory, locomotory flow should occur whenever we experience full sensory immersion when on the move. Such attention-demanding locomotion—variously called wayfaring or wayfinding by the many writers drawn to the skill†—is an indispensable facet of hunter-gatherer life today: many anthropological studies of traditional cultures‡ remark on the constant attentiveness and responsiveness to the subtlest of cues displayed by both hunters and gatherers when on the move. This attitude is born of necessity: whether in search of a fleeing animal or of edible plants, you have to be able to accurately read the land (or sky, or sea, or ice) to know where to go, and how to get home afterward. As a result, the navigational and tracking skills demonstrated by traditional hunter-gatherers can be breathtaking. For example, Hugh Brody of the University of the Fraser Valley, British Columbia, reported that the Dane-zaa, or Beaver Indians, can tell the sex of a moose from the angle of its urine trail in the snow, and Claudio Aporta of Dalhousie University, Nova Scotia, found in his studies of the Inuit of the Canadian Arctic that what appears to us as a featureless polar desert is for them a readily discerned meshwork of trails that date back generations, whose details are never written down but passed verbally or by shared experience from individual to individual.

There are few activities for which the dopamine system is more appropriate than wayfaring. All the boxes are checked—exploration, learning, motivation, attention, and awareness—and while we're not yet able to test the link with dopamine function directly (fMRI scanners are not the most portable of devices!), anthropologists invariably find that wayfaring is a deeply significant activity for traditional peoples, and must therefore have a powerful positive psychological impact. For example, Lye Tuck-Po of the University of Science in Malaysia found during her time with the Batek people of Malaysia that their treks through the forest are much more than simply ways of getting from A to B. For one, these journeys are the principal means by which

* Why this should be so intoxicating isn't known for sure. A love for the sensation of falling certainly seems to fly in the face of natural selection. However, a fall may be a wonderfully cheap way of getting from A to B, as long as it ends well. This might even be a psychological holdover from our arboreal past, when an excessive fear of falling would have been a serious drawback.

† Two notable examples are Rebecca Solnit and Tim Ingold, who explore wayfaring in their respective books *Wanderlust* and *Lines: A Brief History*. Many of the anthropological studies I mention here are discussed more thoroughly in the latter book.

‡ These include James Weiner's reports on the Foi of Papua New Guinea, Heonik Kwon's work on the Orochon of Siberia, and the other studies I reference later.

they acquire knowledge, the pursuit of which she describes as a "deeply satisfying source of meaning" for them. Their forest walks are also the main focus of their storytelling.* As a result, the trails through the landscape are imbued with the history of the people who made and used (and continue to use) them. Indeed, Lye was once told that when the Batek feel a longing for their ancestors, they will seek out the old trails that those ancestors made, so that they may walk in their footsteps, relive their wayfaring experiences, and so remember them and learn from them. In short, for the Batek (and the story is much the same for other hunter-gatherer cultures), wayfaring is an absolutely essential part of their lives, through which they develop their individual identities, preserve their culture, and find a profound connection to their homeland, their past, and their people.

Given the obvious necessity for wayfaring in today's hunter-gatherer cultures, we can be quite sure that this activity played a major role in the lives of our own hunter-gatherer ancestors. As such, we might expect it to have left its mark on our present-day psychology. A few candidate mental traits might make a lot of sense when viewed in this light. Consider storytelling again: stories, quite apart from their navigational utility, only work because we're good at extracting meaning from a sequence of events—allowing what comes before or after to color a given mental image. That's exactly what we do when wayfaring: we extract navigational meaning from a sequence of vistas. Could our love of a good story thus ultimately derive from the way our brain has been designed for locomotion? Viewed in this light, our love of music may also be something of a side effect of our locomotory psychological design: split into isolated sounds, a tune does nothing for us, but knit those sounds together in a certain sequence and suddenly we can, again, extract powerful emotional meaning. Aside from the act of navigation itself, wayfaring also requires an ability to visualize distant goals, and hold on to them in the face of setbacks and hardships. As Bernd Heinrich points out in his book *Why We Run*, it's not hard to see the wider applicability of such staying power. And, of course, the exploratory heart of wayfaring now finds expression in

* For traditional cultures, journey stories may serve a deeper function than simple entertainment. Michelle Scalise Sugiyama of the University of Oregon thinks that many of the more fantastical tales—most famously the songlines of the Australian aboriginal Dreamtime—are essentially verbal maps that provide information about the best paths through a landscape, along with advice about foraging, potential hazards, or places to shelter. Landmarks are often personified as petrified once-living humans or giants—a plot device that makes the navigational instructions more memorable by tapping into our highly developed social intelligence.

our general curiosity and excitement in the new. Small wonder we turn again and again to the language of locomotion when speaking of our lives and achievements.

But what of the act of exploratory locomotion itself? Sadly, for all its subsidiary gifts, wayfaring itself is an almost-forgotten art, despite the fact that it would appear to be an easy way (certainly far easier than ultramarathons and extreme sports) of getting a dopamine-flavored dose of *Funktionslust*. For too many journeys these days, our minds—once full participants in the act of locomotion—have become mere passengers in our bodies, and the activity that was the ancestral source of joy has been transformed into a gray, tedious slog. What has happened?

## TWO LEGS BAD, FOUR WHEELS GOOD

The blame for the disembodiment of our mind from the act of locomotion must lay squarely at the door of certain aspects of our culture, which didn't just create our brave new world but, by insulating us from the effects of natural selection, also made sure that our mind could never fully adapt to it. Agriculture obviously looms large here. By reducing our need to wander, farming downgraded the survival importance of geographical knowledge—a trend that has now reached its climax, with most of us having nothing whatsoever to do with food production. There's not a lot that can be done about that as things stand—we can't just go back to being hunter-gatherers, even if we were willing to ditch the many benefits of civilization, for at our current population level most of us would starve. However, there's something else that drives a wedge between our mind and our body—something that we might be able to address, if we have the courage and the imagination for it. Here the impact on our locomotory lives is more direct, but ironically the primary driver of the change was our desire for movement.

As you may have guessed, I'm speaking of the cultural evolution of our locomotory technologies—a trend that's dominated the last few thousand years of our history almost as much as the evolution of natural locomotion dominated the previous 4 billion. One major strand of this trend was the opening up of those forms of movement—aquatic, aerial, orbital—for which we were poorly adapted. But that's only one side of the story. While we had terrestrial locomotion covered by our biological adaptations (and covered rather well), there were other animals that could do it faster, and once we'd worked out how to domesticate horses, we jumped at the chance to make use of their higher speeds. From that moment, our path to locomotory disenfranchisement was set. It was horses that made wheels useful, and wheels that allowed the internal combustion engine to be pressed into locomotory service, giving us greater and greater performance with less and less effort. Therein lies the trap, you see, for over the last 4 billion years, one directive has dominated locomotory adaptation more than any other, and that is the push for maximum efficiency, a.k.a. the "get as much as possible for as little as possible" principle. Our motor learning is clearly founded on this precept: it's why we all ended up walking so efficiently. How ironic that the same impetus now all but stops us walking at all.

Cars and other personal motor vehicles make a pretty easy target for anyone concerned about where humanity is headed. For one, they're directly responsible for a truly appalling amount of death. The World Health Organization recorded 1.24 million road traffic deaths in 2010 (and that didn't include China). To put that in terms that are easier to grasp, that's equivalent to at least ten medium-size aircraft crashing every day, with no survivors. It's the eighth-highest cause of death worldwide and the top killer of fifteen- to twenty-nine-year-olds—and that's before we've considered the indirect loss of life caused by cars' contribution to the obesity crisis, air pollution, and climate change. Our collective tolerance of such a ghastly toll defies belief. Every bit as troubling, however, is the insidious impact of motorized transport on our relationship with the spaces in which we live, and with our own bodies. Through such influences, our obsession with powered locomotion is not only taking our lives away; it's also making our lives not worth living. Worse, as we'll see shortly, our collective desire has all the trappings of a society-wide addiction that's proving harder and harder to kick.

At the heart of this problem lies the impact of motorized transport on wayfaring. Now, it's not impossible to do this by car. Diana Young of the University of Queensland has studied aboriginal Australian drivers whose approach to the activity is more or less the same as their approach to walking or running: they navigate on the hoof, and experience full conscious sensory immersion as they go. Unfortunately, there are few environments that could support such a free-and-easy driving ethos (though you wouldn't know it looking at SUV ads): either the off-road terrain is inappropriate for wheeled transport or the population density is too high, making untrammeled motoring unsafe or unworkable. Roads (and railroads) are the obvious answer to this problem. Of course, such structures had been around for a long time before motorized transport was invented, but rather than enabling wayfaring, they began to bring about its demise. Successful navigation by road requires only that you know how to choose the right road. Once on it, you can more or less switch off—as long as you don't wander off the path, you'll eventually reach your intended destination. Thanks to the invention of carriages and, later, trains, a person now doesn't even need to be conscious to go somewhere, which if you think about it, defeats the evolutionary object of having consciousness in the first place!

Travel by train or other large-scale transportation systems is thus the very opposite of wayfaring, giving a traveler little sense of active exploration. One can use trains to discover new places, of course, but the discovery won't generally be made while on board. Gazing out of the window at the landscape racing by (assuming the landscape is visible), therapeutic as it may be, is a poor substitute for the sensory immersion that occurs when you're making your own way through that landscape. Also, aside from choosing which train to board, a traveler has almost no locomotory agency—he or she can't decide exactly when to set off, and is completely powerless should anything go wrong (often to the extent of not even knowing what's gone wrong). Few indeed are the passengers who haven't experienced the intense frustration engendered by such situations.

This, of course, is what makes cars so irresistible. They appear to offer the best of both worlds: all the speed and effortlessness of trains coupled with the individual freedom of self-generated locomotion (as the marketing divisions of car manufacturers know full well). Sometimes they come close to fulfilling this promise, but it's a rare treat. One problem is that the more people turn to cars, the more choked the roads become, landing us right back where we started: stripped of much of our locomotory agency. And the purported freedom of car travel is something of an illusion anyway. The choice of roads is nominally much wider than when going by train, but given our obsession with getting to our destination as quickly as possible, the alternative routes quickly evaporate. When travel is just a way of getting from A to B, it's difficult to shake the notion that travel time is anything other than wasted time. Then again, given the crushing dreariness of so many car journeys, one can hardly blame us for holding this opinion. The required level of conscious sensory engagement with the actual process of locomotion is so minimal that drivers may arrive at their destination with no memory of the journey at all. With satnavs the sensory deprivation is nearly complete—we just go where we're told, when we're told. Then there are a whole load of additional drawbacks that don't bother the train traveler, such as the trouble of finding a place to park or the paradoxical requirement of full concentration from the driver despite the unchanging nature of the surroundings or the static configuration of his or her body (a bit of wistful gazing out of the window and you're likely to find yourself contributing to the year's 1.24 million).

All of that wouldn't be so bad if we were more sparing in our car use. Trains may reduce their passengers to powerless packages, but their reach is limited, and once we disembark we become the masters of our locomotory fate once more. These days, cars can go almost everywhere that we'd want to go, with the result that growing numbers of us barely do any locomotion at all that's not powered by an internal combustion engine. Tragically, it's becoming ever more difficult to resist this latest and darkest episode of our locomotory evolution. Our ever-expanding cities and towns are now all too often expressly designed for the car to the exclusion of the pedestrian, as is reflected, for instance, in the enormous nonwalkable distances that can separate people's homes from their places of work and, worse, that separate their homes from their friends' and family's homes. In the United Kingdom, for example, a recent study found that the average distance between someone's home and his or her place of work was 9.3 miles and increasing.* In the United States the equivalent figure, when it was last assessed in 2001, was 12.1 miles.† Apart from making a complete mockery of the supposed benefits of higher transport speeds, this increasing geographical fragmentation of our lives is disintegrating communities and turning us into a society of loners. It doesn't help that what's considered a nonwalkable distance appears to be shrinking: the more we get used to our four-wheeled enhancements, the feebler our own bodies seem to be. A two-hour commute is hardly uncommon these days, but the idea of walking six miles to and from work is nearly unthinkable, despite the fact that the journey would probably take about the same time.‡

The litany of noxious influences continues: as cars take over our shared spaces, it's becoming increasingly difficult and dangerous for children to play outside, which gives rise to all kinds of detrimental effects on their physical, social, and psychological development. Small wonder that learning to drive

---

* Office of National Statistics, UK, "2011 Census Analysis—Distance Travelled to Work," 2014, available at http://www.ons.gov.uk/ons/rel/census/2011-census-analysis/distance-travelled-to-work/index.html.

† P. S. Hu and T. R. Reuscher, "Summary of Travel Trends: 2001 National Household Travel Survey," Washington, DC: US Department of Transportation, Federal Highway Administration, 2004.

‡ Of course, many may be reluctant to walk for fear of violent crime, but the perceived and actual risk would be substantially reduced if more people were to go pedestrian.

is now considered to be such an important rite of passage—being carless is tantamount to having no locomotory capability at all. This is why I liken our current car-obsessed state to a society-wide drug addiction: what was once a luxury item has become a near necessity as the spatial nodes of our lives have become increasingly far-flung—giving up cars now would lead to difficult withdrawal symptoms—and while the meager psychological highs diminish as traffic density increases, the negative social, physical, and mental side effects keep getting worse and worse. Not for nothing did J. R. R. Tolkien curse the invention of the "infernal combustion engine."[§] This is nothing less than a full-scale assault on the human spirit.

So, what can we do about it? The first bit of good news is that many of us are at least vaguely aware that something is deeply wrong. Drivers' surveys are finding fairly high levels of discontent, often coupled with the sincere wish to use cars less. Relatively few respondents belong to the die-hard, driving-is-my-right category. On a related note, many of us are becoming uncomfortably aware of the fact that we should be doing more exercise, and are signing up for gym membership to make up the shortfall. This is, of course, much better than doing no exercise at all, but one wonders what our ancestors would say if they could see us on our treadmills, with our senses disengaged, our exquisite locomotory skills employed to go nowhere at all. Then there's the rise of tourism, which in many of its manifestations allows us to temporarily fulfill our desire to discover a place through self-generated locomotion.

Of course, wanting to ditch the car and being able to ditch the car are two very different things when we're forced by the car-centric design of our settlements to travel long distances on a daily basis. Bicycles present the clearest and cleanest alternative, being moderately fast while maintaining or even increasing locomotory agency, all using nothing more than good old-fashioned legwork. They can also be coupled to certain modes of public transport, should travel distances seem prohibitively long. The potential benefits of widespread cycle use are enormous: a recent (2014) report by Rachel Aldred of the University of Westminster, commissioned by British Cycling, found that an increase in UK bicycle use to Danish levels would knock £17 billion off the nation's health-care bill, decrease road deaths by 30 percent, save four hundred productive life years, thanks to the reduced air pollution, and interestingly, increase retail sales by a quarter. Unfortunately, there are formidable obstacles in the path of such a wholesale change of behavior. The

§This he did in a letter to his son Christopher on April 30, 1944.

biggest problem is that as long as cars dominate the roads, going by bike will often seem intolerably perilous: it isn't much fun to pedal alongside 1- to 3-ton, fume-belching lumps of speeding metal whose pilots often view you as a frustrating obstruction. Protected bike lanes are an obviously good place to start, but are taking an exasperatingly long time to catch on, being common in only a few enlightened countries, such as the Netherlands.

Therein lies the main barrier to the breaking of our collective car addiction. It's not something we can do alone, because the viability of a car-free life (or even a reduced-car life) is so heavily dependent on the design of our settlements and the paths that link them, over which we have almost no influence. Although the rise of widespread car ownership is a recent phenomenon—only about sixty years old—infrastructural change has been so rapid that it's now going to take major sustained political will to reverse the trend. That said, many of us are fortunate enough to live in a democracy, in which a powerful enough collective will can sometimes translate into societal change. But do we care enough? Sixty years may not be long in the grand scheme of things, but it doesn't take much for us to become habituated to a way of doing things, so most of us now blithely accept the dreadful consequences of car addiction as an unavoidable part of modern life. That's why consensus-challenging stunts such as that dreamed up by Austrian engineer Hermann Knoflacher may be our best hope. He created the walkmobile—a simple car-size wooden frame that a person can wear and walk (even cycle) around with. Since the 1970s Knoflacher has used this simple device to shine a spotlight on the enormous amount of space that a car demands, thus unmasking the anti-social heart of private car ownership. Another heartening phenomenon is the recent rise of so-called Critical Mass bike rides, when large groups of cyclists take to the roads to temporarily reclaim the streets and shake the car-obsessed status quo. These events now occur in over three hundred cities worldwide. In some places, the tide appears to be turning: Amsterdam, Copenhagen, and Bogotá are notable for their cycle-friendly designs, and plans are afoot in Helsinki to implement a smartphone-coordinated multimode public transport system so effective that private cars in the city could become obsolete by 2025. If such projects succeed in generating sufficient worldwide momentum (so to speak), we might just find that the car's stranglehold on modern society can be loosened, and that the path back to real locomotory freedom—and a life more in tune with our biological heritage—is still open to us.

Should leg power end up winning out over engine power, it will probably still be a long time before society manages to kick its car addiction.

Nevertheless, even in these gasoline-soaked times, it doesn't take much to rediscover something of our ancestral locomotory joy. In the delightful *Wanderlust*, Rebecca Solnit writes that one doesn't have to be in the wilderness to wayfare—cities can be ripe with opportunities for engaged, immersive locomotion, too. Simply discarding the notion of destinations and time limits for a while does wonders for the sense of discovery. Without deadlines and route maps, we can afford to be fully present as we wander—to properly notice our surroundings rather than letting everything unrelated to our narrow spatial goals pass us by. Our mind becomes embodied once more. As we start to pay heed to the nuances of our environment, drinking in the contours and textures and aromas, we may come to connect more deeply with the places through which we wander. And with every step we take, we write our history in the trails we leave behind, however subtle, as did those who came before us. Whenever we retread those paths as wayfarers, we cannot help but relive the embodied experiences of the people that made them. While it's unlikely that we'll ever fully rediscover the depth of connection that characterizes hunter-gatherers (wayfaring that's motivated by genuine survival need will always mean more than its recreational counterpart), we might find that the simple act of movement acquires something of the power it once had to draw us closer to our world, to our rich past, and to ourselves.

----------

The idea of reenacting past journeys takes on special significance when considered in the light of the evolution of locomotion, thanks to the influence of self-generated motion on the process of natural selection. Tim Ingold, in his *Lines: A Brief History*, laments the passing of the Lamarckian view of life, which incorrectly regarded the inheritance of acquired characteristics as the driving force of evolution. The reason for Ingold's dismay is that, according to certain readings of the opposing Darwinian view, what an organism does or learns has no relevance beyond its own life span. This view can lend the process of evolution a fractured, sterile air, for it treats organisms as temporally self-contained, with only their genes jumping from generation to generation, as if (using Ingold's words) an organism "exists solely to be itself. . . . It neither carries forward the life-course of its antecedents nor anticipates that of its descendants." But that's not what evolution is like, thanks to the central importance of locomotion in its unfolding: what an organism does or learns to do—neither of which are fully specified by its genes—can have a staggering impact on the future evolution of its lineage, for the simple reason that locomotion can take it to new environments and thus expose it to a whole host of new selective pressures.

With this in mind, consider again the evolutionary narrative of the human lineage from its slime-propelled prokaryote infancy. Recall the ancient adoption of an amoeboid crawling way of life or the origin of cilia and their later takeover by the nervous system in the early days of the animal kingdom. Then there was the internal reorganization of the cytoskeleton to make muscle, which first took over steering, and later, propulsion; alongside these developments came the breakage of radial symmetry and the redeployment of repeated developmental modules along the primary axis rather than around it—all in connection with crawling and the rediscovery of Mexican waves in a muscular context—that heralded the coming of the Bilateria. While the various progenitors of the animal phyla initiated their respective body plan experiments that marked the Cambrian explosion, the origin of our own chordate phylum (once we'd flipped upside down) was marked by the appearance of the notochord and the undulatory method of aquatic propulsion, later refined by the overlaying of the vertebral column and the growth of paired fins. Many millions of years later, the first breath of air set in motion a sequence of events that would in time see the ossification of the vertebral column and fin skeletons, the adaptation of the latter for underwater walking, and the move to land. In our line, the energetic efficiency of four-legged walking and running was gradually improved (at the cost of stability) with the shift from a sprawling to an upright posture, before our ancestors took to the trees. Our more recent primate ancestors then grew particularly fond of the bounty to be gleaned at the precarious tips of branches, leading to the evolution of opposable thumbs and big toes. With the adoption of orthograde postures by apes, the evolutionary fates of the fore- and hindlimbs could be more resolutely sundered (thanks to the old Hox genes), which in the hominin line led us back down to the ground and to the later refinement of our ancestors' terrestrial skills, particularly bipedal running. That opened the door to ever-more-successful hunting forays, which likely led to the final expansion of our brain, and (in a different sense) to the expansion of our species to every corner of the globe. To cap it all off, you and I appeared, having inherited the consciousness and exploratory desire to contemplate the whole wondrous 4-billion-year history.

Looking back on that grand narrative, it's difficult to pick out a single chapter that wasn't initiated by a creature's movements. The immediate ancestors of the chordates were surely undulating before the notochord evolved, the pretetrapods crawling on river beds before their fins became limbs, and the preprimates sniffing around fine branches before their hands and feet became fully prehensile. Only through such adventures and their physical consequences did the selective pressures become manifest that would eventually bring about the anatomical adaptations. The process is strongly reminiscent of the way we humans learn to walk—we have to try it out first, and only later do we become good at it. And just as our mastery of walking opens up all kinds of new learning possibilities, both locomotory and nonlocomotory, so, too, do the anatomical responses to the choices made by an ancestor unlock further pathways for adventure in its descendants. The undulations that would eventually cause the appearance of the notochord could only have appeared in a bilaterian body, which itself could only have arisen following the adoption of crawling by an earlier cnidarian-grade animal. Without a doubt (and paraphrasing a certain fictionalized Roman general), what a creature does in the course of its life may truly echo through time.

The influence of an organism's wanderings on its descendants means that our bodies and minds are far more than collected documents pertaining to the nature of our ancestors in anatomical and psychological form. Each of us is a living embodiment of the countless journeys made by those same restless ancestors. I know of no more beautiful testament to the deep connection we share with the rest of the living world. The question we must now ask ourselves is—are we really willing to disregard that heritage in favor of our filthy, deadly locomotory technologies? Our 4-billion-year history of self-generated movement has given us everything we hold dear, down to the very awareness and curiosity that's enabled us to piece that history together. It's now up to us to decide whether that long, wondrous story ends here.

# *Acknowledgments*

If you will forgive me for using yet another locomotory metaphor, this has been a long road, but never a lonely one. From the very beginning, I've been blessed with the support and guidance of many generous-spirited people who have helped me on my way. I am extremely grateful for their companionship, whether occasional or lasting. First, I extend my heartfelt thanks to my wonderful agent, Peter Tallack. Without his advice and encouragement early on, or indeed his willingness to take a chance on an untried author, this book would never have got off the ground (sorry—there's another one!). His unfailing enthusiasm has been invaluable ever since. I am also tremendously grateful to my editors at Basic Books for their constructive advice and patient support: first Tisse Takagi and later Quynh Do, who gamely took over when Tisse moved on. My thanks also to my production editor at Basic Books, Melissa Veronesi, and copyeditor, Iris Bass, for their support, care, and expertise in the later stages of the project, and, last but not least, to T. J. Kelleher, for filling in during the editorial gap, and for his backing from the outset.

My story frequently took me far outside my scientific comfort zone (such as it is), so I am extremely grateful to the many experts, both here in Cambridge and elsewhere, who have graciously given of their time and imparted their wisdom to keep me pointing in the right direction. My special thanks go to George Lauder, Karen Adolph, Xu Xing, and their postdoc and research students, all of whom warmly welcomed me into their labs in Harvard, New York, and Beijing, and also to Holger Babinsky, Robin Crompton, Michael Akam, Simon Laughlin, Paul Maderson, Mike Brooke, Jenny Clack, Tim Smithson, Robert Dudley, Henry Disney, and most of all to Adrian Friday, whose scintillating lectures and fascinating tea-break conversations have been inspiring me for years. I am grateful to Gáspár Jékely, Ulrike Müller, Eric Tytell, David DeRosier, Dennis Thomas, Peter Southwood, Scott Zona, David Hanke, Simon Conway Morris, Jean-Bernard Caron, Günter Bechly, Guy Carpenter, David Midgley, Gil Iosilevskii, David Hone, Rob Theodore, and Russell Stebbings for providing or otherwise helping to obtain some of the images. Special thanks go to Coral Bailey for her beautiful illustrations. I would also like to thank Charlie Ellington and Dave Unwin, who long ago set me on the path on which I now find myself, when they agreed that a PhD on pterodactyl flight might just work.

Finally, I thank my family and friends for keeping me going through the tricky bits, and for indulging my locomotory preoccupation over the last couple of years. I couldn't have done it without them.

# *List of Illustrations*

# *Bibliography*

GENERAL

Alexander, R. McN. 1968. *Animal Mechanics*. London: Sidgwick & Jackson.

———. 2003. *Principles of Animal Locomotion*. Princeton, NJ, and Oxford: Princeton University Press.

Arthur, W. 2014. *Evolving Animals: The Story of Our Kingdom*. Cambridge, UK: Cambridge University Press.

Barnes, R. S. K., P. Calow, P. J. W. Olive, D. W. Golding, and J. I. Spicer. 2001. *The Invertebrates: A Synthesis*. 3rd ed. Oxford: Blackwell Science Ltd.

Biewener, A. A. 2003. *Animal Locomotion*. New York: Oxford University Press.

Clark, R. B. 1964. *Dynamics in Metazoan Evolution*. Oxford: Clarendon Press.

Dawkins, R. 2004. *The Ancestor's Tale*. London: Weidenfeld & Nicolson Ltd.

Denny, M., and A. McFadzean. 2011. *Engineering Animals: How Life Works*. Cambridge, MA: Harvard University Press.

Gray, J. 1968. *Animal Locomotion*. London: Weidenfeld & Nicolson Ltd.

Heinrich, B. 2001. *Why We Run: A Natural History*. New York: HarperCollins.

Lane, N. 2009. *Life Ascending*. London: Profile Books Ltd.

Marey, E. J. 1874. *Animal Mechanism: A Treatise on Terrestrial and Aerial Locomotion*. London: Henry S. King & Company.

McDougall, C. 2009. *Born to Run*. New York: Alfred A. Knopf.

Shubin, N. H. 2008. *Your Inner Fish*. London: Allen Lane.

Vogel, S. 1988. *Life's Devices*. Princeton, NJ: Princeton University Press.

———. 2013. *Comparative Biomechanics: Life's Physical World*. 2nd ed. Princeton, NJ: Princeton University Press.

CHAPTER 1: JUST PUT ONE FOOT IN FRONT OF THE OTHER

Aiello, L. C., and P. Wheeler. 1995. "The Expensive-Tissue Hypothesis." *Current Anthropology* 36 (2): 199–221.

Alexander, R. McN. 1984. "Walking and Running." *American Scientist* 72: 348–54.

———. 1991. "Energy-Saving Mechanisms in Walking and Running." *Journal of Experimental Biology* 160: 55–69.

Bartlett, J. L., and R. Kram. 2008. "Changing the Demand on Specific Muscle Groups Affects the Walk-Run Transition Speed." *Journal of Experimental Biology* 211 (Pt 8): 1281–88.

Bramble, D. M., and D. E. Lieberman. 2004. "Endurance Running and the Evolution of *Homo*." *Nature* 432 (7015): 345–52.

Cappellini, G. 2006. "Motor Patterns in Human Walking and Running." *Journal of Neurophysiology* 95 (6): 3426–37.

Cunningham, C. B., N. Schilling, C. Anders, and D. R. Carrier. 2010. "The Influence of Foot Posture on the Cost of Transport in Humans." *Journal of Experimental Biology* 213 (5): 790–97.

Lieberman, D. E. 2012. "What We Can Learn About Running from Barefoot Running." *Exercise and Sport Sciences Reviews* 40 (2): 63–72.

———. 2013. *The Story of the Human Body*. New York: Pantheon Books.

McGeer, T. 1993. "Dynamics and Control of Bipedal Locomotion." *Journal of Theoretical Biology* 163 (3): 277–314.

Minetti, A. E. 1998. "The Biomechanics of Skipping Gaits: A Third Locomotion Paradigm?" *Proceedings of the Royal Society B: Biological Sciences* 265 (1402): 1227–33.

Muybridge, E. 1877. *Animal Locomotion: An Electro-Photographic Investigation of Consecutive Phases of Animal Movements*. Philadelphia: Photogravure Co.

Ruxton, G. D., and D. M. Wilkinson. 2013. "Endurance Running and Its Relevance to Scavenging by Early Hominins." *Evolution* 67 (3): 861–67.

CHAPTER 2: TWO LEGS GOOD

Almécija, S., D. M. Alba, and S. Moyà-Solà. 2012. "The Thumb of Miocene Apes: New Insights from Castell de Barberà (Catalonia, Spain)." *American Journal of Physical Anthropology* 148 (3): 436–50.

Bloch, J. I. 2002. "Grasping Primate Origins." *Science* 298 (5598): 1606–10.

Cartmill, M. 1974. "Rethinking Primate Origins." *Science* 184 (4135): 436–43.

Cartmill, M., P. Lemelin, and D. Schmitt. 2007. "Primate Gaits and Primate Origins." In *Primate Origins: Adaptations and Evolution*, edited by M. J. Ravosa and M. Dagosto, 403–35. Boston, MA: Springer US.

Crompton, R. H., T. C. Pataky, R. Savage, K. D'Août, M. R. Bennett, M. H. Day, K. Bates, S. Morse, and W. I. Sellers. 2012. "Human-like External Function of the Foot, and Fully Upright Gait, Confirmed in the 3.66 Million Year Old Laetoli Hominin Footprints by Topographic Statistics, Experimental Footprint-Formation and Computer Simulation." *Journal of the Royal Society Interface* 9 (69): 707–19.

Crompton, R. H., W. I. Sellers, and S. K. S. Thorpe. 2010. "Arboreality, Terrestriality and Bipedalism." *Philosophical Transactions of the Royal Society B: Biological Sciences* 365 (1556): 3301–14.

Hunt, K. D. 1991. "Mechanical Implications of Chimpanzee Positional Behavior." *American Journal of Physical Anthropology* 86 (4): 521–36.

Johanson, D. C., and M. Taieb. 1976. "Plio—Pleistocene Hominid Discoveries in Hadar, Ethiopia." *Nature* 260 (5549): 293–97.

Kivell, T. L., and D. Schmitt. 2009. "Independent Evolution of Knuckle-Walking in African Apes Shows That Humans Did Not Evolve from a Knuckle-Walking Ancestor." *Proceedings of the National Academy of Sciences of the United States of America* 106 (34): 14241–46.

Leakey, M. D., and R. L. Hay. 1979. "Pliocene Footprints in the Laetolil Beds at Laetoli, Northern Tanzania." *Nature* 278 (5702): 317–23.

Lemelin, P., and D. Schmitt. 2007. "Origins of Grasping and Locomotor Adaptations in Primates: Comparative and Experimental Approaches Using an Opossum Model." In *Primate Origins: Adaptations and Evolution*, edited by M. J. Ravosa and M. Dagosto, 329–80. Boston, MA: Springer US.

Lovejoy, C. O., B. Latimer, G. Suwa, B. Asfaw, and T. D. White. 2009. "Combining Prehension and Propulsion: The Foot of *Ardipithecus ramidus*." *Science* 326 (5949): 72, 72e1–72e8.

Lovejoy, C. O., S. W. Simpson, T. D. White, B. Asfaw, and G. Suwa. 2009. "Careful Climbing in the Miocene: The Forelimbs of *Ardipithecus ramidus* and Humans Are Primitive." *Science* 326 (5949): 70, 70e1–70e8.

Lovejoy, C. O., G. Suwa, L. Spurlock, B. Asfaw, and T. D. White. 2009. "The Pelvis and Femur of *Ardipithecus ramidus*: The Emergence of Upright Walking." *Science* 326 (5949): 71–71, 71e1–71e6.

Moyà-Solà, S. 2004. "*Pierolapithecus catalaunicus*, a New Middle Miocene Great Ape from Spain." *Science* 306 (5700): 1339–44.

Nagano, A., B. R. Umberger, M. W. Marzke, and K. G. M. Gerritsen. 2005. "Neuromusculoskeletal Computer Modeling and Simulation of Upright, Straight-Legged, Bipedal Locomotion of *Australopithecus afarensis* (A.L. 288–1)." *American Journal of Physical Anthropology* 126 (1): 2–13.

Niemitz, C. 2010. "The Evolution of the Upright Posture and Gait—a Review and a New Synthesis." *Naturwissenschaften* 97 (3): 241–63.

Raichlen, D. A., A. D. Gordon, W. E. H. Harcourt-Smith, A. D. Foster, and Wm. Randall Haas. 2010. "Laetoli Footprints Preserve Earliest Direct Evidence of

Human-Like Bipedal Biomechanics." Edited by Karen Rosenberg. *PLoS ONE* 5 (3): e9769.

Reed, K. E. 2008. "Paleoecological Patterns at the Hadar Hominin Site, Afar Regional State, Ethiopia." *Journal of Human Evolution* 54 (6): 743–68.

Sockol, M. D., D. A. Raichlen, and H. Pontzer. 2007. "Chimpanzee Locomotor Energetics and the Origin of Human Bipedalism." *Proceedings of the National Academy of Sciences* 104 (30): 12265–69.

Soligo, C., and R. D. Martin. 2006. "Adaptive Origins of Primates Revisited." *Journal of Human Evolution* 50 (4): 414–30.

Stanford, C. B. 2003. *Upright: The Evolutionary Key to Becoming Human*. Boston: Houghton Mifflin Harcourt.

Thorpe, S. K. S., R. L. Holder, and R. H. Crompton. 2007. "Origin of Human Bipedalism As an Adaptation for Locomotion on Flexible Branches." *Science* 316 (5829): 1328–31.

Wang, W., R. H. Crompton, T. S. Carey, M. M. Günther, Y. Li, R. Savage, and W. I. Sellers. 2004. "Comparison of Inverse-Dynamics Musculo-Skeletal Models of AL 288–1 *Australopithecus afarensis* and KNM-WT 15000 *Homo ergaster* to Modern Humans, with Implications for the Evolution of Bipedalism." *Journal of Human Evolution* 47 (6): 453–78.

## CHAPTER 3: LEAPS OF FAITH

Alexander, D. E., E.-P. Gong, L. D. Martin, D. A. Burnham, and A. R. Falk. 2010. "Model Tests of Gliding with Different Hindwing Configurations in the Four-Winged Dromaeosaurid *Microraptor gui*." *Proceedings of the National Academy of Sciences* 107 (7): 2972–76.

Babinsky, H. 2003. "How Do Wings Work?" *Physics Education* 38 (6): 497–503.

Boyce, C. K., C. L. Hotton, M. L. Fogel, G. D. Cody, R. M. Hazen, A. H. Knoll, and F. M. Hueber. 2007. "Devonian Landscape Heterogeneity Recorded by a Giant Fungus." *Geology* 35 (5): 399–402.

Crummer, C. A. 2013. "Aerodynamics at the Particle Level." arXiv:nlin.CD/0507032v10.

Dial, K. P. 2003. "Wing-Assisted Incline Running and the Evolution of Flight." *Science* 299 (5605): 402–4.

Dial, R. 2004. "The Distribution of Free Space and Its Relation to Canopy Composition at Six Forest Sites." *Forest Science* 50 (3): 312–25.

Dudley, R., G. Byrnes, S. P. Yanoviak, B. Borrell, R. M. Brown, and J. A. McGuire. 2007. "Gliding and the Functional Origins of Flight: Biomechanical Novelty or Necessity?" *Annual Review of Ecology, Evolution, and Systematics* 38 (1): 179–201.

Emerson, S. B., and M. A. R. Koehl. 1990. "The Interaction of Behavioral and

Morphological Change in the Evolution of a Novel Locomotor Type: 'Flying' Frogs." *Source: Evolution* 44 (8): 1931–46.

Engel, M. S., and D. A. Grimaldi. 2004. "New Light Shed on the Oldest Insect." *Nature* 427 (6975): 627–30.

Lingham-Soliar, T., A. Feduccia, and X. Wang. 2007. "A New Chinese Specimen Indicates That 'Protofeathers' in the Early Cretaceous Theropod Dinosaur *Sinosauropteryx* Are Degraded Collagen Fibres." *Proceedings of the Royal Society B: Biological Sciences* 274 (1620): 1823–29.

Maderson, P. F. A., W. J. Hillenius, U. Hiller, and C. C. Dove. 2009. "Towards a Comprehensive Model of Feather Regeneration." *Journal of Morphology* 270 (10): 1166–1208.

Ostrom, J. H. 1975. "The Origin of Birds." *Annual Review of Earth and Planetary Sciences* 3: 55–77.

Prum, R. O., and A. H. Brush. 2002. "The Evolutionary Origin and Diversification of Feathers." *Quarterly Review of Biology* 77 (3): 261–95.

Rayner, J. M. V. 1985. "Cursorial Gliding in Protobirds: An Expanded Version of a Discussion Contribution." In *The Beginnings of Birds*, edited by M. K. Hecht, J. H. Ostrom, G. Viohl, and P. Wellnhofer, 289–92. Eichstätt: Freunde des Jura-Museum.

Wilkinson, M. T. 2007. "Sailing the Skies: The Improbable Aeronautical Success of the Pterosaurs." *Journal of Experimental Biology* 210 (10): 1663–71.

Wootton, R. J., and C. P. Ellington. 1991. "Biomechanics and the Origin of Insect Flight." In *Biomechanics in Evolution*, edited by J. M. V. Rayner and R. J. Wootton, 99–112. Cambridge, UK: Cambridge University Press.

Xu, X., X. Zheng, C. Sullivan, X. Wang, L. Xing, Y. Wang, X. Zhang, J. K. O'Connor, F. Zhang, and Y. Pan. 2015. "A Bizarre Jurassic Maniraptoran Theropod with Preserved Evidence of Membranous Wings." *Nature* 521 (7550): 70–73.

Xu, X., Z. H. Zhou, and X. Wang. 2000. "The Smallest Known Non-Avian Theropod Dinosaur." *Nature* 408 (6813): 705–8.

Xu, X., Z. Zhou, X. Wang, X. Kuang, F. Zhang, and X. Du. 2003. "Four-Winged Dinosaurs from China." *Nature* 421 (6921): 335–40.

Yanoviak, S. P., Y. Munk, M. Kaspari, and R. Dudley. 2010. "Aerial Manoeuvrability in Wingless Gliding Ants (*Cephalotes atratus*)." *Proceedings of the Royal Society B: Biological Sciences* 277 (1691): 2199–2204.

Zhang, F., S. L. Kearns, P. J. Orr, M. J. Benton, Z. Zhou, D. Johnson, X. Xu, and X. Wang. 2010. "Fossilized Melanosomes and the Colour of Cretaceous Dinosaurs and Birds." *Nature* 463 (7284): 1075–78.

Zheng, X., Z. Zhou, X. Wang, F. Zhang, X. Zhang, Y. Wang, G. Wei, S. Wang, and X. Xu. 2013. "Hind Wings in Basal Birds and the Evolution of Leg Feathers." *Science* 339 (6125): 1309–12.

CHAPTER 4: GO-FASTER STRIPE

Bainbridge, R. 1958. "The Speed of Swimming of Fish as Related to Size and to the Frequency and Amplitude of the Tail Beat." *Journal of Experimental Biology* 35: 109–33.

Conway Morris, S., and J.-B. Caron. 2012. "*Pikaia gracilens* Walcott, a Stem-Group Chordate from the Middle Cambrian of British Columbia." *Biological Reviews* 87 (2): 480–512.

Dabiri, J. O. 2009. "Optimal Vortex Formation as a Unifying Principle in Biological Propulsion." *Annual Review of Fluid Mechanics* 41 (1): 17–33.

Gray, J. 1933. "Studies in Animal Locomotion. I. The Movement of Fish with Special Reference to the Eel." *Journal of Experimental Biology* 10: 88–104.

Lacalli, T. 2012. "The Middle Cambrian Fossil *Pikaia* and the Evolution of Chordate Swimming." *EvoDevo* 3 (1): 12.

Lauder, G. V., and E. G. Drucker. 2002. "Forces, Fishes, and Fluids: Hydrodynamic Mechanisms of Aquatic Locomotion." *News in Physiological Sciences* 17 (6): 235–40.

Müller, U. K., and J. L. van Leeuwen. 2006. "Undulatory Fish Swimming: From Muscles to Flow." *Fish and Fisheries* 7 (2): 84–103.

Stott, R. 2012. *Darwin's Ghosts*. London: Bloomsbury Publishing.

Taylor, G. K., R. L. Nudds, and A. L. R. Thomas. 2003. "Flying and Swimming Animals Cruise at a Strouhal Number Tuned for High Power Efficiency." *Nature* 425 (6959): 707–11.

Tytell, E. D. 2004. "The Hydrodynamics of Eel Swimming: I. Wake Structure." *Journal of Experimental Biology* 207 (11): 1825–41.

———. 2007. "Do Trout Swim Better than Eels? Challenges for Estimating Performance Based on the Wake of Self-Propelled Bodies." *Experiments in Fluids* 43. Berlin, Heidelberg: Springer Berlin Heidelberg: 701–12.

Tytell, E. D., I. Borazjani, F. Sotiropoulos, T. V. Baker, E. J. Anderson, and G. V. Lauder. 2010. "Disentangling the Functional Roles of Morphology and Motion in the Swimming of Fish." *Integrative and Comparative Biology* 50 (6): 1140–54.

Van Leeuwen, J. L. 1999. "A Mechanical Analysis of Myomere Shape in Fish." *Journal of Experimental Biology* 202 (23): 3405–14.

Webb, P. W. 1982. "Locomotor Patterns in the Evolution of Actinopterygian Fishes." *Integrative and Comparative Biology* 22 (2): 329–42.

CHAPTER 5: THE IMPROBABLE INVASION

Clark, J. A. 2012. *Gaining Ground: The Origin and Evolution of Tetrapods*. 2nd ed. Bloomington: Indiana University Press.

Coates, M. I., and J. A. Clack. 1990. "Polydactyly in the Earliest Known Tetrapod Limbs." *Nature* 347 (September): 66–69.

———. 1991. "Fish-like Gills and Breathing in the Earliest Known Tetrapod." *Nature* 352 (July): 234–36.

Daniels, C. B., S. Orgeig, L. C. Sullivan, N. Ling, M. B. Bennett, S. Schürch, A. L. Val, and C. J. Brauner. 2004. "The Origin and Evolution of the Surfactant System in Fish: Insights into the Evolution of Lungs and Swim Bladders." *Physiological and Biochemical Zoology* 77 (5): 732–49.

Fish, F. E., and L. D. Shannahan. 2000. "The Role of the Pectoral Fins in Body Trim of Sharks." *Journal of Fish Biology* 56 (5): 1062–73.

Fricke, H., and K. Hissmann. 1992. "Locomotion, Fin Coordination and Body Form of the Living Coelacanth *Latimeria chalumnae*." *Environmental Biology of Fishes* 34 (4): 329–56.

Geoffroy Saint-Hilaire, E. 1802. "Description d'un Nouveau Genre de Poisson." *Annales Du Musée d'Histoire Naturelle Paris* 1: 57–68.

Gibb, A. C., M. A. Ashley-Ross, and S. T. Hsieh. 2013. "Thrash, Flip, or Jump: The Behavioral and Functional Continuum of Terrestrial Locomotion in Teleost Fishes." *Integrative and Comparative Biology* 53 (2): 295–306.

Graham, J. B., and H. J. Lee. 2004. "Breathing Air in Air: In What Ways Might Extant Amphibious Fish Biology Relate to Prevailing Concepts about Early Tetrapods, the Evolution of Vertebrate Air Breathing, and the Vertebrate Land Transition?" *Physiological and Biochemical Zoology* 77 (5): 720–31.

Lauder, G. V. 2000. "Function of the Caudal Fin During Locomotion in Fishes: Kinematics, Flow Visualization, and Evolutionary Patterns." *American Zoologist* 40 (1): 101–22.

Laurin, M. 2010. *How Vertebrates Left the Water*. Berkeley: University of California Press.

Longo, S., M. Riccio, and A. R. McCune. 2013. "Homology of Lungs and Gas Bladders: Insights from Arterial Vasculature." *Journal of Morphology* 274 (6): 687–703.

Perry, S. F., R. J. A. Wilson, C. Straus, M. B. Harris, and J. E. Remmers. 2001. "Which Came First, the Lung or the Breath?" *Comparative Biochemistry and Physiology Part A: Molecular & Integrative Physiology* 129 (1): 37–47.

Pierce, S. E., J. R. Hutchinson, and J. A. Clack. 2013. "Historical Perspectives on the Evolution of Tetrapodomorph Movement." *Integrative and Comparative Biology* 53 (2): 209–23.

Rewcastle, S. C. 1981. "Stance and Gait in Tetrapods: An Evolutionary Scenario." *Symposia of the Zoological Society of London* 48: 239–67.

Romer, A. S. 1955. "Herpetichthyes, Amphibioidei, Choanichthyes or Sarcopterygii?" *Nature* 176 (4472): 126–27.

———. 1958. "Tetrapod Limbs and Early Tetrapod Life." *Evolution* 12 (3): 365–69.

Wilga, C. D., and G. V. Lauder. 2000. "Three-Dimensional Kinematics and Wake Structure of the Pectoral Fins during Locomotion in Leopard Sharks *Triakis semifasciata*." *Journal of Experimental Biology* 203 (15): 2261–78.

———. 2002. "Function of the Heterocercal Tail in Sharks: Quantitative Wake Dynamics during Steady Horizontal Swimming and Vertical Maneuvering." *Journal of Experimental Biology* 205 (16): 2365–74.

Zimmer, C. 1998. *At the Water's Edge*. New York: Free Press.

CHAPTER 6: A WINNING FORMULA

Adamska, M., S. M. Degnan, K. M. Green, M. Adamski, A. Craigie, C. Larroux, and B. M. Degnan. 2007. "Wnt and TGF-β Expression in the Sponge *Amphimedon queenslandica* and the Origin of Metazoan Embryonic Patterning." Edited by James Fraser. *PLoS ONE* 2 (10): e1031.

Arthur, W. 2011. *Evolution: A Developmental Approach*. Chichester: Wiley-Blackwell.

Ball, E. E., D. M. de Jong, B. Schierwater, C. Shinzato, D. C. Hayward, and D. J. Miller. 2007. "Implications of Cnidarian Gene Expression Patterns for the Origins of Bilaterality—Is the Glass Half Full or Half Empty?" *Integrative and Comparative Biology* 47 (5): 701–11.

Boero, F., B. Schierwater, and S. Piraino. 2007. "Cnidarian Milestones in Metazoan Evolution." *Integrative and Comparative Biology* 47 (5): 693–700.

Carroll, S. B. 2005. *Endless Forms Most Beautiful*. New York: W. W. Norton & Company.

Coates, M. I., and M. J. Cohn. 1998. "Fins, Limbs, and Tails: Outgrowths and Axial Patterning in Vertebrate Evolution." *BioEssays* 20 (5): 371–81.

Couso, J. P. 2009. "Segmentation, Metamerism and the Cambrian Explosion." *International Journal of Developmental Biology* 53: 1305–16.

Gilbert, Scott F. 2013. *Developmental Biology*. 10th ed. Sunderland, MA: Sinauer Associates, Inc.

Gould, S. J. 1989. *Wonderful Life*. New York: W. W. Norton & Company.

Haszprunar, G., and A. Anninger. 2000. "Molluscan Muscle Systems in Development and Evolution." *Journal of Zoological Systematics and Evolutionary Research* 38 (3): 157–63.

Hejnol, A., and M. Q. Martindale. 2008. "Acoel Development Supports a Simple Planula-like Urbilaterian." *Philosophical Transactions of the Royal Society B: Biological Sciences* 363 (1496): 1493–1501.

Holley, S. A., P. D. Jackson, Y. Sasai, B. Lu, E. M. de Robertis, F. M. Hoffmann, and E. L. Ferguson. 1995. "A Conserved System for Dorsal-Ventral Patterning in Insects and Vertebrates Involving Sog and Chordin." *Nature* 376 (6537): 249–53.

Hughes, C. L., and T. C. Kaufman. 2002. "Exploring Myriapod Segmentation: The Expression Patterns of Even-Skipped, Engrailed, and Wingless in a Centipede." *Developmental Biology* 247 (1): 47–61.

Knoll, A. H. 1999. "Early Animal Evolution: Emerging Views from Comparative

Biology and Geology." *Science* 284 (5423): 2129–37.

Manuel, M. 2009. "Early Evolution of Symmetry and Polarity in Metazoan Body Plans." *Comptes Rendus Biologies* 332 (2–3): 184–209.

Meinhardt, H. 2002. "The Radial-Symmetric Hydra and the Evolution of the Bilateral Body Plan: An Old Body Became a Young Brain." *BioEssays* 24 (2): 185–91.

Minelli, A., and G. Fusco. 2004. "Evo-Devo Perspectives on Segmentation: Model Organisms, and Beyond." *Trends in Ecology & Evolution* 19 (8): 423–29.

Niehrs, C. 2010. "On Growth and Form: A Cartesian Coordinate System of Wnt and BMP Signaling Specifies Bilaterian Body Axes." *Development* 137 (6): 845–57.

Piraino, S., G. Zega, C. di Benedetto, A. Leone, A. Dell'Anna, R. Pennati, D. Candia Carnevali, V. Schmid, and H. Reichert. 2011. "Complex Neural Architecture in the Diploblastic Larva of *Clava multicornis* (Hydrozoa, Cnidaria)." *Journal of Comparative Neurology* 519 (10): 1931–51.

Pourquié, O. 2003. "The Segmentation Clock: Converting Embryonic Time into Spatial Pattern." *Science* 301 (5631): 328–30.

Pueyo, J. I., R. Lanfear, and J. P. Couso. 2008. "Ancestral Notch-Mediated Segmentation Revealed in the Cockroach *Periplaneta americana*." *Proceedings of the National Academy of Sciences* 105 (43): 16614–19.

Stollewerk, A., M. Schoppmeier, and W. G. M. Damen. 2003. "Involvement of Notch and Delta Genes in Spider Segmentation." *Nature* 423 (6942): 863–65.

CHAPTER 7: BRAIN AND BRAWN

Arendt, D., A. S. Denes, G. Jékely, and K. Tessmar-Raible. 2008. "The Evolution of Nervous System Centralization." *Philosophical Transactions of the Royal Society B: Biological Sciences* 363 (1496): 1523–28.

Ashcroft, F. 2012. *The Spark of Life*. London: Allen Lane.

Coates, M. M. 2003. "Visual Ecology and Functional Morphology of Cubozoa (Cnidaria)." *Integrative and Comparative Biology* 43 (4): 542–48.

Costello, J. H., S. P. Colin, and J. O. Dabiri. 2008. "Medusan Morphospace: Phylogenetic Constraints, Biomechanical Solutions, and Ecological Consequences." *Invertebrate Biology* 127 (3): 265–90.

Elgeti, J., and G. Gompper. 2013. "Emergence of Metachronal Waves in Cilia Arrays." *Proceedings of the National Academy of Sciences* 110 (12): 4470–75.

Ellwanger, K., A. Eich, and M. Nickel. 2007. "GABA and Glutamate Specifically Induce Contractions in the Sponge *Tethya wilhelma*." *Journal of Comparative Physiology A* 193 (1): 1–11.

Jékely, G. 2011. "Origin and Early Evolution of Neural Circuits for the Control of Ciliary Locomotion." *Proceedings of the Royal Society B: Biological Sciences* 278 (1707): 914–22.

Jékely, G., J. Colombelli, H. Hausen, K. Guy, E. Stelzer, F. Nédélec, and D. Arendt. 2008. "Mechanism of Phototaxis in Marine Zooplankton." *Nature* 456 (7220): 395–99.

Leys, S. P. 2015. "Elements of a 'Nervous System' in Sponges." *Journal of Experimental Biology* 218 (4): 581–91.

Ludeman, D. A., N. Farrar, A. Riesgo, J. Paps, and S. P. Leys. 2014. "Evolutionary Origins of Sensation in Metazoans: Functional Evidence for a New Sensory Organ in Sponges." *BMC Evolutionary Biology* 14 (1): 3.

Nickel, M. 2010. "Evolutionary Emergence of Synaptic Nervous Systems: What Can We Learn from the Non-Synaptic, Nerveless Porifera?" *Invertebrate Biology* 129 (1): 1–16.

Nickel, M., C. Scheer, J. U. Hammel, J. Herzen, and F. Beckmann. 2011. "The Contractile Sponge Epithelium *Sensu Lato*-Body Contraction of the Demosponge *Tethya wilhelma* Is Mediated by the Pinacoderm." *Journal of Experimental Biology* 214 (10): 1692–98.

Satterlie, R. A. 2002. "Neuronal Control of Swimming in Jellyfish: A Comparative Story." *Canadian Journal of Zoology* 80 (10): 1654–69.

Satterlie, R. A., and T. G. Nolen. 2001. "Why Do Cubomedusae Have Only Four Swim Pacemakers?" *Journal of Experimental Biology* 204: 1413–19.

Simmons, P. J., and D. Young. 2010. *Nerve Cells and Animal Behaviour*. 3rd ed. Cambridge, UK: Cambridge University Press.

Stöckl, A. L., R. Petie, and D.-E. Nilsson. 2011. "Setting the Pace: New Insights into Central Pattern Generator Interactions in Box Jellyfish Swimming." *PLoS ONE* 6 (11): e27201.

Tosches, M. A., and D. Arendt. 2013. "The Bilaterian Forebrain: An Evolutionary Chimaera." *Current Opinion in Neurobiology* 23 (6). Elsevier Ltd: 1080–89.

CHAPTER 8: GIVE IT A REST

Azuma, A., and Y. Okuno. 1987. "Flight of a Samara, *Alsomitra macrocarpa*." *Journal of Theoretical Biology* 129 (3): 263–74.

Berge, J., O. Varpe, M. A. Moline, A. Wold, P. E. Renaud, M. Daase, and S. Falk-Petersen. 2012. "Retention of Ice-Associated Amphipods: Possible Consequences for an Ice-Free Arctic Ocean." *Biology Letters* 8 (6): 1012–15.

Berner, R. A. 2006. "GEOCARBSULF: A Combined Model for Phanerozoic Atmospheric $O_2$ and $CO_2$." *Geochimica et Cosmochimica Acta* 70 (23): 5653–64.

Christian, K. A., R. V. Baudinette, and Y. Pamula. 1997. "Energetic Costs of Activity by Lizards in the Field." *Functional Ecology* 11: 392–97.

Cleveland, L. R., and A. V. Grimstone. 1964. "The Fine Structure of the Flagellate *Mixotricha Paradoxa* and Its Associated Micro-Organisms." *Proceedings of the Royal Society B: Biological Sciences* 159 (977): 668–86.

Cronberg, N. 2006. "Microarthropods Mediate Sperm Transfer in Mosses." *Science* 313 (5791): 1255–1255.

Dawkins, R. 1982. *The Extended Phenotype*. Oxford: Oxford University Press.

Evert, R. F., and S. E. Eichhorn. 2013. *Biology of Plants*. 8th ed. New York: W. H. Freeman.

Floudas, D., M. Binder, R. Riley, K. Barry, R. A. Blanchette, B. Henrissat, A. T. Martinez, et al. 2012. "The Paleozoic Origin of Enzymatic Lignin Decomposition Reconstructed from 31 Fungal Genomes." *Science* 336 (6089): 1715–19.

Forterre, Y., J. M. Skotheim, J. Dumais, and L. Mahadevan. 2005. "How the Venus Flytrap Snaps." *Nature* 433 (7024): 421–25.

Fulcher, B. A., and P. J. Motta. 2006. "Suction Disk Performance of Echeneid Fishes." *Canadian Journal of Zoology* 84 (1): 42–50.

Hodick, D., and A. Sievers. 1988. "The Action Potential of *Dionaea muscipula* Ellis." *Planta* 174 (1): 8–18.

Iosilevskii, G., and D. Weihs. 2009. "Hydrodynamics of Sailing of the Portuguese Man-of-War *Physalia physalis*." *Journal of the Royal Society Interface* 6 (36): 613–26.

Klavins, S. D., D. W. Kellogg, M. Krings, E. L. Taylor, and T. N. Taylor. 2005. "Coprolites in a Middle Triassic Cycad Pollen Cone: Evidence for Insect Pollination in Early Cycads?" *Evolutionary Ecology Research* 7 (3): 479–88.

Mapstone, G. M. 2014. "Global Diversity and Review of Siphonophorae (Cnidaria: Hydrozoa)." *PLoS ONE* 9 (2): e87737.

Marmottant, P., A. Ponomarenko, and D. Bienaime. 2013. "The Walk and Jump of *Equisetum* Spores." *Proceedings of the Royal Society B: Biological Sciences* 280 (1770): 20131465.

Moore, H., K. Dvoráková, N. Jenkins, and W. Breed. 2002. "Exceptional Sperm Cooperation in the Wood Mouse." *Nature* 418 (6894): 174–77.

Moore, J. D., and E. R. Trueman. 1971. "Swimming of the Scallop, *Chlamys opercularis* (L.)." *Journal of Experimental Marine Biology and Ecology* 6 (1936): 179–85.

Pontzer, H., and R. W. Wrangham. 2004. "Climbing and the Daily Energy Cost of Locomotion in Wild Chimpanzees: Implications for Hominoid Locomotor Evolution." *Journal of Human Evolution* 46 (3): 315–33.

Rosenstiel, T. N., E. E. Shortlidge, A. N. Melnychenko, J. F. Pankow, and S. M. Eppley. 2012. "Sex-Specific Volatile Compounds Influence Microarthropod-Mediated Fertilization of Moss." *Nature* 489 (7416): 431–33.

Sallan, L. C., and M. I. Coates. 2010. "End-Devonian Extinction and a Bottleneck in the Early Evolution of Modern Jawed Vertebrates." *Proceedings of the National Academy of Sciences* 107 (22): 10131–10135.

Tiffney, B. H. 2004. "Vertebrate Dispersal of Seed Plants Through Time." *Annual Review of Ecology, Evolution, and Systematics* 35 (1): 1–29.

Trewavas, A. 2003. "Aspects of Plant Intelligence." *Annals of Botany* 92 (1): 1–20.

Varshney, K., S. Chang, and Z. J. Wang. 2012. "The Kinematics of Falling Maple Seeds and the Initial Transition to a Helical Motion." *Nonlinearity* 25 (1): C1–8.

Weiblen, George D. 2002. "How to Be a Fig Wasp." *Annual Review of Entomology* 47 (1): 299–330.

CHAPTER 9: EXODUS

Blanchoin, L., R. Boujemaa-Paterski, C. Sykes, and J. Plastino. 2014. "Actin Dynamics, Architecture, and Mechanics in Cell Motility." *Physiological Reviews* 94 (1): 235–63.

Brumley, D. R., K. Y. Wan, M. Polin, and R. E. Goldstein. 2014. "Flagellar Synchronization through Direct Hydrodynamic Interactions." *eLife* 3: e02750.

Cascales, E., R. Lloubès, and J. N. Sturgis. 2008. "The TolQ-TolR Proteins Energize TolA and Share Homologies with the Flagellar Motor Proteins MotA-MotB." *Molecular Microbiology* 42 (3): 795–807.

Cooper, R. M. 2012. "The Origins of Directional Persistence in Amoeboid Motility." PhD thesis: Princeton University.

Dobell, C. 1932. *Antony van Leeuwenhoek and His "Little Animals."* New York: Harcourt, Brace and Company.

Gibbons, B. H., and I. R. Gibbons. 1973. "The Effect of Partial Extraction of Dynein Arms on the Movement of Reactivated Sea-Urchin Sperm." *Journal of Cell Science* 13 (2): 337–57.

Hoffmann, P. M. 2012. *Life's Ratchet*. New York: Basic Books.

Insall, R. H. 2010. "Understanding Eukaryotic Chemotaxis: A Pseudopod-Centred View." *Nature Reviews Molecular Cell Biology* 11 (6): 453–58.

Jarrell, K. F., and S.-V. Albers. 2012. "The Archaellum: An Old Motility Structure with a New Name." *Trends in Microbiology* 20 (7): 307–12.

Jarrell, K. F., and M. J. McBride. 2008. "The Surprisingly Diverse Ways That Prokaryotes Move." *Nature Reviews Microbiology* 6 (6): 466–76.

Jékely, G. 2009. "Evolution of Phototaxis." *Philosophical Transactions of the Royal Society B: Biological Sciences* 364 (1531): 2795–2808.

Jékely, G., and D. Arendt. 2006. "Evolution of Intraflagellar Transport from Coated Vesicles and Autogenous Origin of the Eukaryotic Cilium." *BioEssays* 28 (2): 191–98.

Lindemann, C. B., and K. A. Lesich. 2010. "Flagellar and Ciliary Beating: The Proven and the Possible." *Journal of Cell Science* 123 (4): 519–28.

Niklas, K. J. 2004. "The Cell Walls That Bind the Tree of Life." *BioScience* 54: 831–41.

Pallen, M. J., and N. J. Matzke. 2006. "From The Origin of Species to the Origin of Bacterial Flagella." *Nature Reviews Microbiology* 4 (10): 784–90.

Peabody, C. R. 2003. "Type II Protein Secretion and Its Relationship to Bacterial Type IV Pili and Archaeal Flagella." *Microbiology* 149 (11): 3051–72.

Satir, P. 1965. "Studies on Cilia: II. Examination of the Distal Region of the Ciliary Shaft and the Role of the Filaments in Motility." *Journal of Cell Biology* 26 (3): 805–34.

Skerker, J. M., and H. C. Berg. 2001. "Direct Observation of Extension and Retraction of Type IV Pili." *Proceedings of the National Academy of Sciences* 98 (12): 6901–4.

Van Leewenhoeck, A. 1674. "More Observations from Mr. Leewenhook, in a Letter of Sept. 7. 1674. Sent to the Publisher." *Philosophical Transactions of the Royal Society of London* 9 (108): 178–82.

———. 1677. "Observations, Communicated to the Publisher by Mr. Antony van Leewenhoeck, in a Dutch Letter of the 9th of Octob. 1676. Here English'd: Concerning Little Animals by Him Observed in Rain-Well-Sea. and Snow Water; as Also in Water Wherein Pepper Had Lain In." *Philosophical Transactions of the Royal Society of London* 12 (133–142): 821–31.

Wang, S., H. Arellano-Santoyo, P. A. Combs, and J. W. Shaevitz. 2010. "Actin-like Cytoskeleton Filaments Contribute to Cell Mechanics in Bacteria." *Proceedings of the National Academy of Sciences* 107 (20): 9182–85.

Wong, T., A. Amidi, A. Dodds, S. Siddiqi, J. Wang, T. Yep, D. G. Tamang, and M. H. Saier. 2007. "Evolution of the Bacterial Flagellum Cumulative Evidence Indicates That Flagella Developed as Modular Systems, with Many Components Deriving from Other Systems." *Microbe* 2 (7): 335–40.

CHAPTER 10: LOCOMOTIVE SOULS

Acredolo, L. P., A. Adams, and S. W. Goodwyn. 1984. "The Role of Self-Produced Movement and Visual Tracking in Infant Spatial Orientation." *Journal of Experimental Child Psychology* 38 (2): 312–27.

Adolph, K. E., W. G. Cole, M. Komati, J. S. Garciaguirre, D. Badaly, J. M. Lingeman, G. L. Y. Chan, and R. B. Sotsky. 2012. "How Do You Learn to Walk? Thousands of Steps and Dozens of Falls per Day." *Psychological Science* 23 (11): 1387–94.

Adolph, K. E., and S. R. Robinson. 2013. "The Road to Walking." In *Oxford Handbook of Developmental Psychology*, edited by P. D. Zelazo, 403–43. New York: Oxford University Press.

Anable, J. 2005. "'Complacent Car Addicts' or 'Aspiring Environmentalists'? Identifying Travel Behaviour Segments Using Attitude Theory." *Transport Policy* 12 (1): 65–78.

Anderson, D. I., J. J. Campos, D. C. Witherington, A. Dahl, M. Rivera, M. He, I. Uchiyama, and M. Barbu-Roth. 2013. "The Role of Locomotion in Psychological Development." *Frontiers in Psychology* 4.

Aporta, C. 2009. "The Trail as Home: Inuit and Their Pan-Arctic Network of Routes." *Human Ecology* 37 (2): 131–46.

Balcombe, J. 2006. *Pleasurable Kingdom*. Basingstoke, New York: Macmillan.

Beeler, J. A., C. R. M. Frazier, and X. Zhuang. 2012. "Putting Desire on a Budget: Dopamine and Energy Expenditure, Reconciling Reward and Resources." *Frontiers in Integrative Neuroscience* 6.

Bertenthal, B. I., J. J. Campos, and K. C. Barrett. 1984. "Self-Produced Locomotion." In *Continuities and Discontinuities in Development*, edited by R. N. Emde and R. J. Harmon, 175–210. New York: Plenum Press.

Böhm, S, C. Jones, C. Land, and M. Paterson, eds. 2006. *Against Automobility*. Oxford: Blackwell Publishing Ltd.

Brody, H. 2002. *Maps and Dreams*. London: Faber & Faber.

Buckley, R. 2012. "Rush as a Key Motivation in Skilled Adventure Tourism: Resolving the Risk Recreation Paradox." *Tourism Management* 33 (4): 961–70.

Bushnell, E. W. 2000. "Two Steps Forward, One Step Back." *Infancy* 1 (2): 225–30.

Campos, J. J., D. I. Anderson, M. Barbu-Roth, E. M. Hubbard, M. J. Hertenstein, and D. Witherington. 2000. "Travel Broadens the Mind." *Infancy* 1 (2): 149–219.

Clearfield, M. W., C. N. Osborne, and M. Mullen. 2008. "Learning by Looking: Infants' Social Looking Behavior across the Transition from Crawling to Walking." *Journal of Experimental Child Psychology* 100 (4): 297–307.

Dietrich, A. 2004. "Endocannabinoids and Exercise." *British Journal of Sports Medicine* 38 (5): 536–41.

Ekkekakis, P., G. Parfitt, and S. J. Petruzzello. 2011. "The Pleasure and Displeasure People Feel When They Exercise at Different Intensities." *Sports Medicine* 41 (8): 641–71.

Gibson, E. 1988. "Exploratory Behavior In The Development Of Perceiving, Acting, And The Acquiring Of Knowledge." *Annual Review of Psychology*.

Hughes, D. P., S. B. Andersen, N. L. Hywel-Jones, W. Himaman, J. Billen, and J. J. Boomsma. 2011. "Behavioral Mechanisms and Morphological Symptoms of Zombie Ants Dying from Fungal Infection." *BMC Ecology* 11 (1): 13.

Ingold, T. 2007. *Lines: A Brief History*. Abingdon, UK: Routledge.

Kringelbach, M. L. 2009. *The Pleasure Center*. New York: Oxford University Press.

Kwon, H. 1998. "The Saddle and the Sledge: Hunting as Comparative Narrative in Siberia and Beyond." *Journal of the Royal Anthropological Institute* 4 (1): 115–27.

Lewis, K. P. 2010. "From Landscapes to Playscapes: The Evolution of Play in Humans and Other Animals." In *The Anthropology of Sport and Human Movement*, edited by R. R. Sands and L. R. Sands, 61–89. Lanham, MD: Lexington.

Lye, T.-P. 2002. "The Significance of Forest to the Emergence of Batek Knowledge in Pahang, Malaysia." *Southeast Asian Studies* 40 (1): 3–22.

Mahler, M. S., F. Pine, and A. Bergman. 1975. *The Psychological Birth Of The Human Infant*. New York: Basic Books.

McAllister, J. E. 2011. "Stuck Fast: A Critical Analysis of the 'New Mobilities' Paradigm." Master's thesis: University of Auckland.

O'Keefe, J. H., R. Vogel, C. J. Lavie, and L. Cordain. 2011. "Exercise Like a Hunter-Gatherer: A Prescription for Organic Physical Fitness." *Progress in Cardiovascular Diseases* 53 (6): 471–79.

Olds, J., and P. Milner. 1954. "Positive Reinforcement Produced by Electrical Stimulation of Septal Area and Other Regions of Rat Brain." *Journal of Comparative and Physiological Psychology* 47 (6): 419–27.

Piaget, J. 1952. *The Origins of Intelligence in Children*. New York: International Universities Press.

Pontoppidan, M.-B., W. Himaman, N. L. Hywel-Jones, J. J. Boomsma, and D. P. Hughes. 2009. "Graveyards on the Move: The Spatio-Temporal Distribution of Dead Ophiocordyceps-Infected Ants." *PLoS ONE* 4 (3): e4835.

Pooley, C. G., D. Horton, G. Scheldeman, C. Mullen, T. Jones, M. Tight, A. Jopson, and A. Chisholm. 2013. "Policies for Promoting Walking and Cycling in England: A View from the Street." *Transport Policy* 27 (May): 66–72.

Raichlen, D. A., A. D. Foster, G. L. Gerdeman, A. Seillier, and A. Giuffrida. 2012. "Wired to Run: Exercise-Induced Endocannabinoid Signaling in Humans and Cursorial Mammals with Implications for the 'Runner's High'." *Journal of Experimental Biology* 215 (8): 1331–36.

Rivière, J., R. Lécuyer, and M. Hickmann. 2009. "Early Locomotion and the Development of Spatial Language: Evidence from Young Children with Motor Impairments." *European Journal of Developmental Psychology* 6 (5): 548–66.

Scalise Sugiyama, M. 2001. "Food, Foragers, and Folklore: The Role of Narrative in Human Subsistence." *Evolution and Human Behavior* 22 (4): 221–40.

Solnit, R. 2000. *Wanderlust: A History of Walking*. New York: Viking Penguin.

Tomasello, M. 1999. *The Cultural Origins of Human Cognition*. Cambridge, MA: Harvard University Press.

Weiner, J. F. 1998. "Revealing the Grounds of Life in Papua New Guinea." *Social Analysis* 42 (3): 135–42.

Young, D. 2001. "The Life and Death of Cars: Private Vehicles on the Pitjantjatjara Lands, South Australia." In *Car Cultures*, edited by D. Miller, 35–57. Oxford: Berg.

Zavestoski, S., and J. Agyeman, eds. 2015. *Incomplete Streets: Processes, Practices, and Possibilities*. Abingdon, UK, and New York: Routledge.

# Index

**Matt Wilkinson** is a zoologist and science writer at the University of Cambridge. His work has been covered in *The Telegraph*, *New Scientist*, and *Nature*. Wilkinson lives in Cambridge, England.